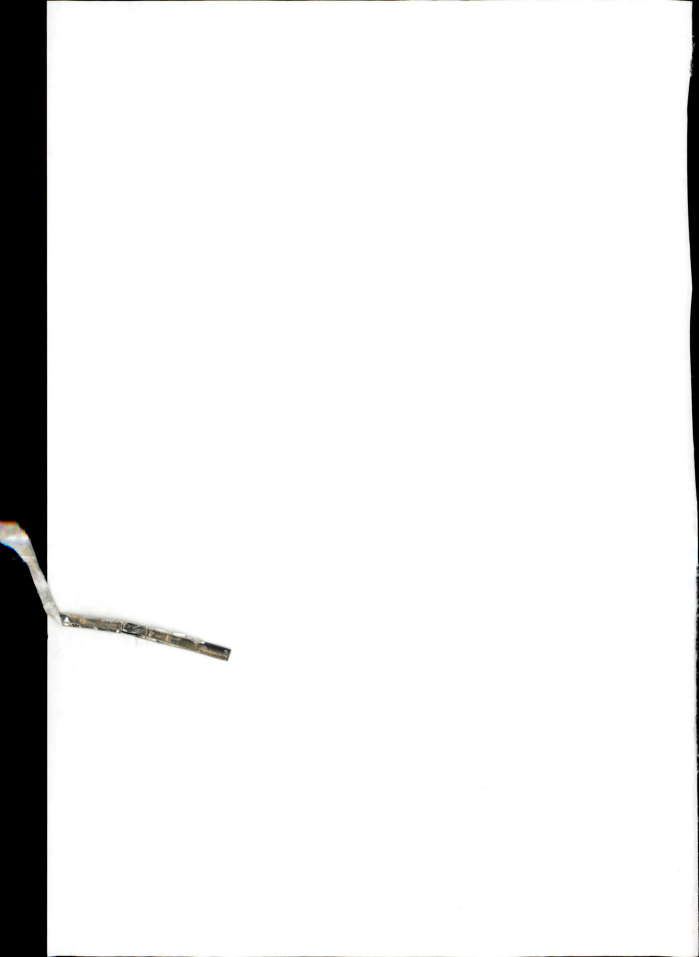

Science Terms Made Easy

A Lexicon of Scientific Words and Their Root Language Origins

Joseph S. Elias

GREENWOOD PRESS
Westport, Connecticut • London

Library of Congress Cataloging-in-Publication Data

Elias, Joseph S., 1948–
  Science terms made easy : a lexicon of scientific words and their root language origins / Joseph S. Elias.
    p. cm.
  Includes bibliographical references and index.
  ISBN 0–313–33896–5 (alk. paper)
  1. Science—Terminology. 2. Latin language—Technical Latin. 3. English language—Etymology. I. Title.
  Q179.E45 2007
  501'.4—dc22   2006026197

British Library Cataloguing in Publication Data is available.

Library of Congress Catalog Card Number: 2006026197
ISBN: 0–313–33896–5

First published in 2007

Greenwood Press, 88 Post Road West, Westport, CT 06881
An imprint of Greenwood Publishing Group, Inc.
www.greenwood.com

Printed in the United States of America

The paper used in this book complies with the
Permanent Paper Standard issued by the National
Information Standards Organization (Z39.48–1984).

10 9 8 7 6 5 4 3 2 1

For all you mean to me,
this book is dedicated to my mother and father;
to my brothers, Edward and Victor;
and to my daughters, Elizabeth, Kate, and Samantha.

Contents

Preface

The idea for writing this book came about as a result of discussions with my pre-service science teacher on effective ways to teach science vocabulary. Years ago, I came to realize that high school and middle school students viewed the complexities of scientific vocabulary as a necessary burden that sometimes interfered with their pursuit of understanding important concepts. Students at these grade levels would complain about words that were unfamiliar or unrecognizable. Quite often science teachers new to the profession would address the vocabulary by developing word lists and definitions or by coming up with simple word association games promoting the ability to recognize words and recall their meanings.

During my years as a teacher of human anatomy and physiology, I developed a student assignment called the "List of 50 Muscles." Students were provided with a list of the muscles, and their task was to examine the names and describe all they could about a given muscle simply by analyzing the name. Muscles such as the pterygoideus internus, the external carpi radialis longus, or my favorite, the sternocleidomastoideus, challenged students to go beyond the words themselves and, in a sense, dissect the word as they would if they were dissecting a preserved specimen. Students discovered that the parts of these scientific terms could be interchanged and still retain their meanings.

As you might imagine, for me as a young teacher, this was a breakthrough of sorts. I became as strong an advocate for inquiry-type teaching approaches to scientific terminology as I was for the teaching of inquiry methods in science itself. I found a way to once again challenge students to think, analyze, and reason their way to a deeper understanding rather than resort to rote memorization.

This, of course, led me to more deeply examine the terminology that I used on a regular basis in all my science classes. I became more curious about the origins and the history of the words. If a student wanted an explanation of a given word, I wanted to be prepared to either point the student in a direction where he could find an answer or, sometimes, to simply tell the tale myself.

What I discovered was that words have histories. They move through cultures and times and mutate along the way. So when you examine the list, you will find descriptions of many roots that will call upon you to make the connections between the original meanings of the roots and their modern counterparts. Sometimes making those connections is a stretch, and you'll have to use your imagination. But through all of this, I found the literal meanings to be simple, if not humble, compared with the rather sophisticated uses of the root words today.

I hope you will value the sidebars. You will probably notice that the ancient Greeks had much to do with science, mathematics, and philosophy. These great thinkers provided the world with its first really grand period of scientific enlightenment. The philosophers of the time pondered the order of the universe. They speculated and hypothesized on all aspects of order and chaos. They spoke of the things that were earthly and of things that were divine, and they used these models as the bases for their perception of the physical world. Many of the terms used in science, especially the physical sciences, have their origins in the Greek language.

The study of living things—anatomy, taxonomy, and medicine—did not really move forward until the next period of scientific enlightenment, in the seventeenth and eighteenth centuries. By that time, the great days of the Greek civilization were long past and the age of exploration and investigation moved more toward Western Europe. Thus, you will notice that many of the root words associated with living things are of Latin rather than Greek origin.

As a final note, this compilation of words is by no means meant to be a complete text of scientific terminology, but it does represent a very healthy collection of the more common words used in science courses in middle and high school science classes. I imagine that students in lower-division college courses will also find this book to be a valuable reference. It is my sincere hope that readers will have as much fun with this compilation of science terminology as I had putting it all together.

Acknowledgments

First and foremost, I wish to acknowledge the many science education students at Kutztown University for their significant contributions to my list of words. Without them, the task of gathering information and developing the final product would have been far more daunting of an undertaking.

I would also like to acknowledge the members of the Department of Secondary Education at Kutztown University. Their support, expertise, guidance, and patience allowed me to focus on the task at hand.

I would also like to thank the regional science teachers who, on occasion, would e-mail or pass along words that caught their interest.

How to Use This Book

I have never underestimated the creativity of teachers. When they were given the right tools and the proper amount of time, the teachers that I have known developed some fascinating perspectives on how to teach science. Virtually all experienced and talented science teachers pride themselves on being able to challenge students to think, reason, predict, hypothesize, and interpret data collected from observation and experimentation. This book provides another valuable component to assist them in their efforts.

Teaching scientific terminology for understanding has always been a challenge for teachers. The words included in this text will provide the teacher with a source for integrating complex terminology into their lessons. I recommend that instructors design activities that call for students to critically examine the words they are learning in ways that encourage them to look deeper into their meanings and historic origins. The sidebars provide historical perspectives and a quick study of interesting people and events that led to the study of science and technology in the modern era. The reader will gain an appreciation of how scientists, mathematicians, and philosophers of past eras were able to develop theories of the order of the universe based on reason rather than experimentation. Many of these theories went unchallenged for over a thousand years.

I would encourage students to become very familiar with the common prefixes and suffixes. Suffixes such as -or and -ion appear repeatedly in words pertaining to actions or processes. Prefixes such as a- or an- and con- or com- are very common in scientific language. If students are made aware of how these word fragments are used, they should be able to recognize their relevance in terms that are new to them. Teachers may also want to point out that the o's have been deliberately removed from many of the word fragments, the reason being that they are generally referred to as "combining vowels." The o is used to connect many commonly used prefixes and suffixes to the root words; such, for instance, is the case with stern-o-mastoid.

This inquiry approach to language not only strengthens the analytical skills of students, it also fosters a sense of independence in the learner. Students quickly learn that they have the power to examine complex words and construct new meanings independently of a teacher or professor.

Abdomen
Latin
abdomen belly, venter
That portion of the body that lies between the lower thorax (chest) and the pelvis.

Abdominalgia
Latin/Greek
abdomen- belly, venter
-algia pain, sense of pain; painful; hurting
Pain in the abdomen; a belly ache.

Abductor
Latin
ab- off, away from
-ducere- to draw or lead
-or a condition or property of things or persons, person that does something
The name given to the function of a skeletal muscle used to pull a body part (arm or leg) away from the midline of the body.

Aberration
Latin
aberrare- deviation from the proper or expected course
-ion state, process, or quality of
The blurring or distortion of an image, typically caused by a defect in the lens.

Abiocoen
Greek
a- without
-bios- life, living organisms or tissue
-coen common, shared
The sum total of the nonliving components of an environment.

Abiotic
Greek
a- without
-bios- life, living organisms or tissue
-ic (ikos) relating to or having some characteristic of
The set of nonliving environmental factors or conditions that are common within a given ecological system.

Abrasion
Latin
abradere- to scrape off
-ion state, process, or quality of
The process of wearing down or scraping off by means of rubbing one object against another object.

Abscess
Greek
ab- off, away from
-cēdere to go
A localized collection of pus in part of the body, formed by tissue disintegration and surrounded by an inflamed area.

Abscission
Latin
ab- off, away from
-caedere- to cut
-ion state, process, or quality of
The shedding of leaves, flowers, or fruits following the formation of the abscission zone.

Absorbance
Latin
ab- off, away from
-sorbere- to suck
-ance brilliance, appearance

The relative ability of the surface of a substance to retain radiant energy.

Abyssal
Greek
a- without
-bussos- bottom
-al of the kind of, pertaining to, having the form or character of
Of or relating to the region of the ocean bottom between the bathyal and hadal zones, from depths of approximately 3,000 to 6,000 meters.

Acanthaceous
Greek/Latin
akanthos- thorn plant
-aceous having the quality of
Resembling or having the quality of the family of plants that bear prickles or spines.

Acanthologist
Greek
akanthos- thorn plant
-logist one who speaks in a certain manner; one who deals with a certain topic
A person who studies spines or spiny creatures.

Acapnia
Greek
a- no, absence of, without, lack of, not
-kapnos smoke, carbon dioxide (CO_2)
A condition marked by the presence of less than the normal amount of CO_2 in blood and tissue.

Acardia
Greek
a- no, absence of, without, lack of, not
-kard- heart, pertaining to the heart
-ia names of diseases, place names, Latinizing plurals
A congenital condition, usually occurring with twins, where one of the two siblings is born without a heart, or a lone heart is shared by the two.

Acaulescent
Latin
a- no, absence of, without, lack of, not
-caulis- stem
-escent being in a specific state, beginning to be
A seemingly stemless plant, though the stem may be small and sometimes belowground.

Accipitrine
Latin
accipiter- hawk
-ine of or relating to
Raptorial, hawklike, belonging to the genus *Accipiter*.

Acclimation
Greek
a- no, absence of, without, lack of, not
-klime- slope
-ion state, process, or quality of
Physiological responses to environmental change.

Accommodation
Latin
ad- to, a direction toward, addition to, near
-commodus- to adjust, suitable
-ion state, process, or quality of
The state or process of adjusting one item to another.

Accuracy
Latin
accuratus- done with care
-cy state, condition, quality
Precision, exactness.

Acetabulum
Latin
aceta- hip
-bul- place for
-um (**singular**) structure
-a (**plural**) structure
Cup-shaped cavity at the base of the hipbone.

Acetylcholine
Latin/Greek
acetum- vinegar
-khole- bile
-ine a chemical substance
A neurotransmitter that mediates the synaptic activity of autonomic synapses and neuromuscular junctions.

Acheiria
Greek
a- no, absence of, without, lack of, not
-chir- hand; pertaining to the hand or hands
-ia names of diseases, place names, or Latinizing plurals
Congenital absence of the hands.

Acidaminuria
Latin
acere- to be sour
-amino- relating to an amine or other compound containing an NH_2 group
-urina urine
A disorder involving the metabolism of protein where excessive amounts of amino acids are found in the urine.

Acidemia
Latin
acere- to be sour
-haima blood

A medical condition in which blood pH is below normal.

Acidic
Latin
acere- to be sour
-ic (ikos) relating to or having some characteristic of
Having the reactions or characteristics of an acid.

Acidiferous
Latin
acere- to be sour
-ferrous bear, carry; produce
Producing or yielding an acid.

Acidize
Latin/Greek
acere- to be sour
-ize to make, to treat, to do something with
To treat with acid.

Acidosis
New Latin
acere- to be sour
-sis action, process, state, condition
The condition in which there is an excessive amount of acid in the blood.

Acoelomate
Latin/Greek
a- no, absence of, without, lack of, not
-coelom- (koilomat) cavity
-ate an organism having these characteristics
An organism lacking a body cavity between the gut and the outer musculature of the body wall.

Acology
Greek
aco- remedy, cure
-logy (logos) used in the names of sciences or bodies of knowledge
The science of remedies; therapeutics.

Acroanesthesia
Greek
acro- outermost; extreme; extremity of the body
-an- without, not
-aisthesis- feeling
-ia names of diseases, place names, or Latinizing plurals
Loss of sensation in the extremities; such as the hands, fingers, toes, and feet.

Acrodendrophile
Greek
acro- high, highest, highest point; top, tip end, outermost; extreme
-dendron- tree, treelike structure
-phile one who loves or has a strong affinity or preference for

In biology, describing a species that lives or thrives in treetop habitats.

Acromegaly
Greek
acro- high, highest, highest point; top, tip end, outermost; extreme
-megas large, big, great
A chronic disease in which the bones of the extremities, face, and jaw become enlarged.

Acrosome
Greek
acro- high, highest, highest point; top, tip end, outermost; extreme
-soma (somatiko) body
A caplike structure at the anterior end of a spermatozoon that produces enzymes aiding in egg penetration.

Actin
Latin
āctus- motion
-inus relating to
A protein found in muscle that, together with myosin, functions in muscle contraction.

Actinoid
Greek
aktin- ray (as of light), radiance, radiating
-oid (oeidēs) resembling; having the appearance of
Having a radial form, as a starfish.

Actinotherapy
Greek
aktin- ray (as of light), radiance, radiating
-therapeuein heal, cure; treatment
Treatment of disease by means of light rays.

Activation
Latin
āctus- to set in motion
-ion state, process, or quality of
Stimulation of activity in an organism or chemical.

Activity
Latin
activus- to drive, do
-ity state of, quality of
The state of being active; energetic action or movement; liveliness.

Actophilous
Greek
acto- seashore, beach
-phile- one who loves or has a strong affinity or preference for
-ous full of, having the quality of, relating to
In biology, organisms thriving on rocky seashores or growing on coasts.

Natural Selection

Over a century ago two men put forth a coherent theory about the origin of new species. The explanation was really quite simple and was based mostly on observations of the natural world. Yet today people in the Western world continue to contest the validity of the theory of evolution based on natural selection.

Charles Darwin and Alfred Russell Wallace contended that the world is full of different species, and that any species, if allowed to do so, will grow at a prolific rate, producing far more progeny than can be handled by its environment. The results are readily observable: the excess population of a given species tends to die off, leaving behind an acceptable number of organisms given the available resources. Darwin believed that the organisms that manage to survive do so because they are best adapted to the particular set of environmental conditions in which they exist. Since survivors tend to live to reproduce, those managing to do so would pass on to the next generation the same or similar genetical traits that allowed them to be among the "selected." And because organisms tend do what comes natural—eat, drink, seek shelter, and breed—the progeny or filial generation would invariably be confronted with environmental stresses influencing their ability to carry out the first three of these natural functions, leading to the imposition of a selective process on their numbers and leaving the survivors to breed among themselves—that is, assuming they are sexual in their habits

Now multiply this process by the time allotted for each generation—which is considerably longer for humans than for rats, for instance. The number of offspring produced by fertile females varies, as does their reproductive viability (how often they reproduce). When we compare the number and frequency of births for rats with those of even more prolific species, such as fleas or bacteria, we naturally find that the more prolific a species is, the greater the likelihood of diversity in genotype and phenotype.

It is all about adaptability. Through selection, over time species tend to become more in tune with their environment. Because of successful adaptation and continual breeding, any given species has the capacity to produce genetic mutations. These continual, chance changes in genetic code over extreme periods of time have the potential of modifying the individuals of a given species to the point to where they significantly differ from their ancestors. These genetically produced modifications are "tested" against environmental conditions and are either selected for or selected against based on whether the organism lives long enough to breed.

Acuminate

Latin
acus- (acuere) to sharpen; needle, point
-ate characterized by having
Describing the tip of some leaves tapering gradually at the end to a point.

Acute

Latin
acus sharp; needle
Severe and sharp, as in pain.

Adactylia

Greek
a- no, absence of, without, lack of, not
-daktulos toe, finger, digit
The absence of digits on the hand or foot.

Adaptation

Latin
ad- to, a direction toward, addition to, near
-aptare- fit, fitted, suited
-ion state, process, or quality of
Modification of an organism or its parts that makes it more fit for existence under the conditions of its environment.

Adduct

Latin
ad- to, a direction toward, addition to, near
-ducere to lead, bring, take, or draw
To draw inward toward the median axis of the body or toward an adjacent part or limb.

Adductor

Latin
ad- to, a direction toward, addition to, near
-ducere- to lead, bring, take, or draw
-or a condition or property of things or persons; person who does something
Any muscle used to draw a body part toward the midline of the body.

Adelopod

Greek
a- no, absence of, without, lack of, not
-delo- visible, clear, clearly seen, obvious
-pod foot
An animal whose feet are not apparent.

Adenalgia
Greek
aden- lymph gland(s)
-algia pain, sense of pain; painful, hurting
A painful swelling in a gland.

Adendric
Greek
a- no, absence of, without, lack of, not
-dendr- tree, resembling a tree
-ic (ikos) relating to or having some characteristic of
Without dendrites.

Adenine
Greek
aden- lymph gland(s)
-ine of or relating to
A white crystalline base found in various animal and vegetable tissues as one of the purine base constituents.

Adenitis
Greek
aden- lymph gland(s)
-itis inflammation, burning
Inflammation of a lymph node or of a gland.

Adenocarcinoma
Greek
aden- lymph gland(s)
-karkinos- crab, cancer
-oma tumor, neoplasm
A malignant tumor originating in glandular epithelium.

Adenofibrosis
Greek/Latin
aden- lymph gland(s)
-fibre- an elongated threadlike structure
-sis action, process, state, condition
Fibroid change in a gland.

Adenoid
Greek
aden- lymph gland(s)
-oid (oeidēs) resembling; having the appearance of
Glandlike lymphoid tissue, similar to the tonsils, located high in the back of the pharynx.

Adenovirus
Greek
aden- lymph gland(s)
-virus poison
Any of a group of DNA-containing viruses that cause conjunctivitis and upper respiratory tract infections in humans.

Adhesive
Latin
ad- to, a direction toward, addition to, near
-haerere- stick to, cling to
-ive performing an action
Tending to cling; sticky.

Adiabatic
Greek
a- no, absence of, without, lack of, not
-diabatos- passable
-ic (ikos) relating to or having some characteristic of
Of, relating to, or being a reversible thermodynamic process that occurs without gain or loss of heat and without a change in entropy.

Adipocyte
Latin
adip- of or pertaining to fat
-cyte (kutos) sac or bladder that contains fluid
A mature fat cell found in animals.

Adiponecrosis
Greek
adip- of or pertaining to fat
-necro- death
-sis action, process, state, condition
Death of fatty tissue occurring in hemorrhagic pancreatitis.

Adipose
Latin
adip- of or pertaining to fat
-ose sugar, carbohydrate
Of a fatty nature; the fat present in the cells of adipose tissue.

Adjuvant
Latin
ad- to, a direction toward, addition to, near
-jungere- to join or unite
-an one that is of, relating to, or belonging to
A substance added to a vaccine to increase its effectiveness.

Adrenal
Latin
ad- to, a direction toward, addition to, near
-ren- the kidneys
-al of the kind of, pertaining to, having the form or character of
Glands located on top of the kidneys.

Advection
Latin
ad- to, a direction toward, addition to, near
-vehere- to carry
-ion state, process, or quality of
The transfer of a property of the atmosphere, such as heat, cold, or humidity, by the horizontal movement of an air mass.

Adventitious
Latin
ad- to, a direction toward, addition to, near
-vent- come
-ous full of, having the quality of, relating to
Describing buds of a plant developing in internodes or on roots.

Adynamandrous
Greek
a- without
-dunamikos- powerful
-androus man, men, male, masculine
Having nonfunctioning male reproductive organs.

Aerenchyma
Latin
aer- air, atmosphere, mist, wind
-enchyma tissue
Large air-filled cells that allow rapid diffusion of oxygen within wetland plants.

Aerobacter
Greek
aer- air, atmosphere, mist, wind
-bacter rod-shaped microorganism
Any genus of bacteria normally found in the intestine.

Aerobic
Greek
aer- air, atmosphere, mist, wind
-bio- life, living organisms or tissue
-ic (ikos) relating to or having some characteristic of
Pertaining to organisms or processes that require the presence of oxygen.

Aerobiont
Greek
aer- air, atmosphere, mist, wind
-bio- life, living organisms or tissue
-ont (einai) to be
Either an organism living in air as distinct from water or soil or an organism requiring oxygen.

Aerolite
Greek
aer- air, atmosphere, mist, wind
-lite- (lith) stone or rock
A meteorite that is composed of a siliceous stony material.

Aerophilous
Greek
aer- air, atmosphere, mist, wind
-phile- one who loves or has a strong affinity or preference for
-ous full of, having the quality of, relating to
Refers to plants that are pollinated by wind or fertilized by airborne pollen.

Aerotaxis
Greek
aer- air, atmosphere, mist, wind
-taxis order or arrangement
Movement of an organism in response to the presence of molecular oxygen.

Affect
Latin
ad- to, a direction toward, addition to, near
-facere to do, carry, bear, bring
To act upon or have an influence upon some behavior.

Affector
Latin
ad- to, a direction toward, addition to, near
-facere- to do, carry, bear, bring
-or a condition or property of things or persons; person who does something
In biology, the term given to a nerve cell.

Afferent
Latin
ad- to, a direction toward, addition to, near
-facere- to do, carry, bear, bring
-ent causing an action, being in a specific state, within
Leading toward a region of interest; carrying toward the center of an organ or section, such as nerves that conduct impulses from the body to the brain or spinal cord.

Agantha
Greek
a- without
-gnatha jaw
A superclass of fish that lack a jaw and a pelvic fin.

Agglutination
a- without
-glutinare- to glue
-ion state, process, or quality of
The process by which red blood cells clump together.

Agonist
Greek
agon- conflict, contest
-ist one who is engaged in
A muscle that is contracting and has an opposing muscle (antagonist) applying force on a bone in the opposite direction.

George Washington Carver

"Our creator is the same and never changes despite the names given Him by people here and in all parts of the world. Even if we gave Him no name at all, He would still be there, within us, waiting to give us good on this earth."

—G.W. Carver

How eloquent this humble man and inventor was during his life. George Washington Carver was born in 1864, near the end of the American Civil War, in Diamond Grove, Missouri. In these troubled times, Carver was kidnapped along with his mother by Confederate night raiders and wound up in Arkansas. Moses Carver, the owner of the farm that was George's birthplace, later found George and reclaimed him. He and his wife, Susan, raised George as their own. His natural mother was never found, and the identity of his father was not known.

He left home at the tender age of 12 to begin his schooling. George suffered all the setbacks associated with racial segregation. He was the first black student ever to be admitted into Simpson College of Indianola, Iowa. There he studied piano and art, but George wanted to study science, so he transferred to Iowa Agricultural College in 1891, when he was 27 years old. George was a diligent student; he earned both a bachelor's and a master's degree in bacterial botany and agriculture in 1897 and became the first black member of the Iowa college.

Later that year, George Washington Carver moved to Tuskegee, Alabama, to become the Director of Agriculture at the Tuskegee Normal and Industrial Institute for Negroes. It was here that Carver began a career that has impacted the lives of millions. He helped revolutionize agricultural practices in the war-torn South. As a result of the continuous planting of either cotton or tobacco, southern plantations had become virtually useless. Carver taught farmers about crop rotation for the purpose of enriching the fields with nutrients. He taught them how to grow peanuts, soybeans, sweet potatoes, and other soil-enriching crops. This brought the South back to life again.

George Washington Carver was never interested in wealth or profit from his work. He lived by his words: "How far you go in life depends on your being tender with the young, compassionate with the aged, sympathetic with the striving, and tolerant of the weak and strong. Because someday in your life you will have been all of these." He held three patents, but he did not patent the numerous discoveries he made while at Tuskegee. He created over 300 products from peanuts and more than 100 products from sweet potatoes.

Carver was a compassionate teacher. He taught his students to love nature and to use the forces of nature for the benefit of all. He believed that education should be "made common" and that all members of the community would profit by an educated society.

George Washington Carver died in 1943. He was honored by President Franklin Roosevelt with a national monument, the first for an African American, near Diamond Grove, the place of his childhood.

Agriculture
Latin
agros- of or belonging to fields or soil
-colere to till
The science, art, and business of cultivating soil, producing crops, and raising livestock; farming.

Agroforestry
Greek/Latin
agros- of or belonging to fields or soil
-foris- outside
-y place for an activity, condition, state
Land management for simultaneous production of food crops and trees.

Aigialophilous
Greek
aigial- beach, seashore, cliff
-phile- one who loves or has a strong affinity or preference for
-ous full of, having the quality of, relating to
A community of organisms that thrive in beach habitats or among pebbles on the beach.

Albedo
Latin
albus- the color white
-oid (oeidēs) resembling; having the appearance of
The ability of the surface of a planet or a moon to reflect light.

Albinism
Latin
albus- the color white
-ism state or condition, quality
The state or condition of being an albino; a group of inherited disorders characterized by deficiency or absence of pigment in the skin, hair, and eyes due to an abnormality in the production of melanin.

Albumin
Latin
albumo- the color white
-in protein or derived from a protein
Blood plasma protein produced in the liver.

Alcohol
Med. Latin from Arabic
al- the
-kuhl- essences obtained by distillation
-ol alcohol
Any of a series of hydroxyl compounds having the general formula $C_nH_{2n+1}OH$.

Aldehyde
Latin
al. dehyd- short for *alcohol dehydrogenate*
Any of a class of highly reactive organic chemical compounds obtained by oxidation of primary alcohols.

Aldosterone
Greek/Latin
al. dehyd- dehydrogenized alcohol
-stereos- solid
-one chemical compound containing oxygen in a carbonyl group
A steroid hormone secreted by the adrenal cortex that regulates the salt and water balance in the body.

Algae (alga)
Latin
alga seaweed
A very large, diverse group of plantlike organisms that are mostly aquatic or marine. They range from the unicellular forms to the extremely large kelp forms.

Algaecide
Latin
alga seaweed
-cide (caedere) to cut, kill, hack at, or strike
Type of pesticide that controls algae in bodies of water.

Algesimeter
Greek
algeis- pain
-meter (metron) instrument or means of measuring; to measure
An instrument used to measure the sensitivity to pain, such as that produced by pricking with a sharp point.

Algesiogenic
Greek
algeis- pain
-gen- to give birth, kind, produce
-ic (ikos) relating to or having some characteristic of
Producing pain.

Alimentary
Latin
alimentum- nourishment, supplying food
-ary of, relating to, or connected with
Pertaining to food or nourishment and to the digestive system/alimentary canal.

Alinasal
Latin/Greek
ala- wing
-nasus- nose
-al of the kind of, pertaining to, having the form or character of
Pertaining to the flaring of the nostrils.

Aliphatic
Greek
aleiphein- to anoint with oil
-ic (ikos) relating to or having some characteristic of
Of or relating to a group of organic chemical compounds with carbon atoms linked in open chains.

Alkalimeter
Latin (from Arabic)/Greek
alkali- (**Latin**) basic (pH more than 7)
alqili- (**Arabic**) ashes (originally from Arabic word *al-qali,* which means "ashes," and recalls the elements Na [sodium] and K [potassium] left in the ashes of burning wood or plants)
-meter (metron) instrument or means of measuring; to measure
An apparatus for measuring concentrations of alkalinity in solutions.

Alkaline
Latin (from Arabic)/Greek
alkali- (**Latin**) basic (pH more than 7)
alqili- (**Arabic**) ashes (originally from Arabic word *al-qali,* which means "ashes," and recalls the elements Na [sodium] and K [potassium] left in the ashes of burning wood or plants)
-ine of or relating to
Relating to or containing the carbonate or hydroxide of an alkali metal (the aqueous solution of which is bitter, slippery, caustic, and basic).

Alkalosis
Latin (from Arabic)/Greek
alkali- (**Latin**) basic (pH more than 7)
alqili- (**Arabic**) ashes (originally from Arabic word *al-qali,* which means "ashes," and recalls the elements Na [sodium] and K [potassium] left in the ashes of burning wood or plants)
-sis action, process, state, condition
The condition in which there is an excessive amount of alkali in the blood.

Alkane
English/Arabic/French
alkyl- (**English**) alcohol
al-kuhl- (**Arabic**) *al-* the + *kuhl* powder of antimony
-(meth)ane an odorless, colorless gas (CH_4)
Any member of the alkane series.

Alkene
Latin (from Arabic)/Greek
alkyl- (**English**) alcohol
al-kuhl- (**Arabic**) *al-* the + *kuhl* powder of antimony
-ene an unsaturated organic compound
Any of a series of unsaturated, open-chain hydrocarbons with one or more carbon-carbon double bonds.

Alkyne
Latin (from Arabic)/Greek
alkyl- (**English**) alcohol
al-kuhl- (**Arabic**) *al-* the + *kuhl* powder of antimony
-ine a chemical compound
Any of a series of open-chain hydrocarbons with a carbon-carbon triple bond.

Allele
Greek
alleion mutually
One of two or more alternative forms of a gene, occupying the same position on paired chromosomes and controlling the same inherited characteristic.

Allergen
Greek
allos- other, different
-gen to give birth, kind, produce
A substance, such as pollen, that causes an allergy.

Alliaceous
Latin
allium- onion, garlic bulb
-aceous having the quality of
Of or pertaining to the botanical genus *Allium*.

Allometry
Greek
allos- other, different
-metria (metron) the process of measuring
The patterns of relationships among structure, function, and size.

Allosaur
Greek
allos- other, different
-sauros lizard
Any one of a group of dinosaurs existing in the late Jurassic and early Cretaceous periods. They had features similar to those of the tyrannosaur, but were small.

Allotropy
Greek
allos- other, different
-trope- bend, curve, turn, a turning; response to a stimulus
-y place for an activity, condition, state
The existence of two or more crystalline or molecular structural forms of an element (rotating light in different directions).

Alloy
Latin
alligare- to bind
-y place for an activity, condition, state
The state of mixing two or more metallic substances where the combination calls for each metal to occupy spaces within the molecules of the other.

Alluvion
Latin (*alluere*)
ad- to, a direction toward, addition to, near
-luere- to wash
-ion state, process, or quality of
The process by which the wash or flow of water inundates a land mass; to wash against.

Altimeter
Latin
altus- high, highest, tall, lofty
-meter (metron) instrument or means of measuring, to measure
A barometer-like device that is used in airplanes to determine altitude.

Altitude
Latin
altus- high, highest, tall, lofty
-ude state, quality, condition of
In astronomy, the angle between an object in the sky and the horizon.

Altricial
Latin
alere- to nourish
-al of the kind of, pertaining to, having the form or character of
Referring to various bird species in which hatchlings are typically weak, naked, and dependent on their parents.

Altruism
Latin
alter- other
-ism state or condition, quality
Instinctive cooperative behavior that is detrimental to the individual but contributes to the survival of the species.

Alveolus
Latin
alveus hollow, belly
Microscopic air-containing sacs in the lungs where gases are exchanged during external respiration.

Amalgam
Greek
a- no, absence of, without, lack of, not
-malgama soft mass
A combination of different elements sometimes mixed with mercury to create an alloy used in dentistry.

Amalgamate
Greek
amalgama- mixture
-ate a derivative of a specific chemical compound or element
To combine or mix a group of elements into an integrated whole; the substance remains a mixture or alloy.

Amblyopia
New Latin
ambly- dull, dim
-optic- eye, optic
-ia names of diseases, place names, or Latinizing plurals
Reduced or dim vision; also called lazy eye.

Ambulacrum
Latin
ambula- walk
-crum planted with trees
-um (**singular**) structure
-a (**plural**) structure
One of the five radial areas on the undersurface of the starfish, from which the tube feet are protruded and withdrawn.

Amictic
Greek
a- no, absence of, without, lack of, not
-miktos- mixed or blended
-ic (ikos) relating to or having some characteristic of
Pertaining to female rotifers, which produce only diploid eggs that cannot be fertilized, or to the eggs produced by such females.

Ammeter
French/Greek
am- (ampere) named for Andre Marie Ampere
-meter (metron) instrument or means of measuring, to measure
A device used to measure electrical current in amperes.

Ammine
Latin
ammonia- a colorless, pungent gas, NH_3
-ine a chemical compound
Any of a class of inorganic coordination compounds of ammonia and a magnetic salt.

Ammophilous
Greek
ammo- sand, sandy beach
-phile- one who loves or has a strong affinity or preference for
-ous full of, having the quality of, relating to
In biology, vegetation that thrives in sandy beach habitats.

Amniocentesis
Greek
amnion- embryo, bowl, lamb
-kentein- to prick, puncture
-sis action, process, state, condition
A surgical procedure in which a small sample of amniotic fluid is drawn from the uterus through a needle inserted in the abdomen.

Amniotic
Greek
amnion- embryo, bowl, lamb
-ic (ikos) relating to or having some characteristic of
Of or relating to the amnion, the sac or fluid that protects the embryo (as in *amniotic sac* or *amniotic fluid*).

Amoeba
Greek
ameibein to change
One-celled aquatic or parasitic organism belonging to the genus *Amoeba*, appearing as a mass of protoplasm with no definite shape.

Amoeboid
Greek
ameibein- to change
-oid (oeidēs) resembling; having the appearance of
Amoeba-like in putting forth pseudopodia.

Amorphous
Greek
a- no, absence of, without, lack of, not
-morph- shape, form, figure, or appearance
-ous full of, having the quality of, relating to
Substance with a disjointed, incomplete crystal lattice or without shape.

Amphibian
Latin
amphi- on both or all sides, around
-bios- life, living organisms or tissue
-an one that is of or relating to or belonging to

An animal capable of living both on land and in water.

Amphibious
Greek
amphi- on both or all sides, around
-bios- life, living organisms or tissue
-ous full of, having the quality of, relating to
Relating to organisms that are able to live both on land and in water.

Amphiboles
Greek
amphi- on both or all sides, around
-bol (ballein) to put or throw
Any of a large group of structurally similar hydrated double-silicate minerals.

Amphigean
Greek
amphi- on both or all sides, around
-ge- earth, world
-an one that is of, relating to, or belonging to
Extending all over the earth, from the equator to both poles.

Amphioxus
Greek
amphi- on both or all sides, around
-oxus sharp
Small, flattened marine organism with a notochord (but no true vertebrae), which gives it a pointed shape; the lancelet.

Amphipathic
Greek
amphi- on both or all sides, around
-path- suffering, disease
-ic (ikos) relating to or having some characteristic of
Relating to protein molecules with one surface containing hydrophilic and the other hydrophobic amino acid residues.

Amphoteric
Greek
ampho- (amphoteros) both, each of two
-ic (ikos) relating to or having some characteristic of
Capable of reacting chemically as either an acid or a base.

Amplitude
Latin
amplus- large, full
-ude state, quality, or condition of
The maximum displacement of wave from a rest position; the measurement of a wave from the normal to the height of the wave (crest) or to the depth of the trough.

Ampulla
Latin
amphi- on both or all sides, around
-phoreus bearer
Any membranous bag shaped like a leathern bottle, as the dilated end of a vessel or duct; especially, the dilations of the semicircular canals of the ear.

Amygdala
Greek
amygdale almond
An almond-shaped region of the brain, located in the medial temporal lobe, believed to play a key role in the emotions.

Amylopsin
Greek
amulon- starch; not ground at a mill
-tripsis- a rubbing (so named by its first being obtained by rubbing a pancreas with glycerin)
-in protein or protein derivative
The starch-digesting amylase produced in the pancreas.

Amyotonia
Greek
a-, ano- no, absence of, lack of, without, not
-myo- muscle
-tonia, -tone tension, pressure
Generalized absence of muscle tone, usually associated with flabby musculature and an increased range in passive movement at joints.

Anabolism
Greek
ana- anew, up
-bol- (ballein) to put or throw
-ism state or condition, quality
Building of complex molecules within a cell.

Anaerobe
Greek
an- no, absence of, without, lack of, not
-aerobe organism requiring oxygen to live
Organism that can live in the absence of atmospheric oxygen.

Analgesic
Greek
an- no, absence of, without, lack of, not
-algesi- pain, sense of pain; painful, hurting
-ic (ikos) relating to or having some characteristic of
Referring to compounds that reduce pain perception.

Analog
Greek
analogos proportionate
In chemistry, a compound in which one or more elements are replaced by other elements.

Claudius Galenus of Pergamum

In the annals of medicine, the writings and teaching of one Claudius Galenus, better known as Galen, overshadow those of any other individual. The medical perspectives of this ancient Greek physician occupied a position of prominence in the training of physicians throughout Europe for over a thousand years. Galen was born in 129 AD in the city of Pergamum, known today as Bergama, Turkey. Like many of the more learned people of his time, he had a wide range of interests, including astronomy, philosophy, astrology, and agriculture. He chose to focus on medicine. After studying medicine in Alexandria and Corinth, he practiced wound treatment in gladiatorial schools.

He moved to Rome, where he began his career as a lecturer and very quickly established himself as an expert in the field. Soon he was appointed physician to the Roman emperor Marcus Aurelius and later to his son Commodus.

Galen found himself in Rome at a time when the Roman Empire was at constant war with factions on its northern border. As the empire slowly crumbled around him, Galen spent his years in Rome doing what he did best, dissecting animals. It was this work that laid the foundation for the practice of medicine for over a thousand years. It wasn't a pretty sight to behold. Galen often dissected live animals, and he would cut certain nerve bundles to observe what happened as a result. Galen was able to identify the causes of paralysis by severing the spinal cords of pigs; he cut the nerve controlling vocalization in the larynx and, of course, discovered that the animal became incapable of making sounds. He noted that blood was carried through vessels, and he made accurate observations about the brain that were contrary to Aristotle's notions of the roles of the brain and the heart in the origination of conscious thought. He had numerous scribes record his observations and draw the organs and blood vessels of the dissected animals, and this resulted in one of the major works based on his research. This seventeen-volume classic was titled *On the Usefulness of the Parts of the Human Body.*

Galen did not, however, do significant work with the human torso. Therefore, centuries later, quite a few of Galen's anatomical drawings proved to be less than accurate, and it became necessary to rob graves and to seek out the bodies of freshly executed prisoners for dissection.

Analysis
Greek
ana- anew, up
-ly- (luein) to loosen, dissolve, dissolution, break
-sis action, process, state, condition
Resolving or separating a whole into its elements or component parts.

Anaphase
Greek
ana- anew, up
-phase a stage
The third of four stages of nuclear division in mitosis and in each of the two divisions of meiosis.

Anastomosis
Greek
ana- anew, up
-stoma- mouth
-sis action, process, state, condition
The connection of separate parts of a branching system to form a network, such as blood vessels.

Anatomy
Greek
ana- anew, up
-temnein to cut

The structure of an animal or plant and any of its parts.

Anconitis
Greek
ancon- elbow
-itis inflammation, burning sensation
An inflammation of the elbow joint.

Androecium
Greek
andros- male
-oikos house
Part of a flower that produces male gametes, or pollen grains.

Androgen
Greek
andros- male
-gen to give birth, kind, produce
Male hormone secreted mostly by the testes and to a lesser amount by the adrenal cortex.

Andronosia
Greek
andros- male
-nosia disease
Diseases occurring most often in males.

Anemia

Greek

an- no, absence of, without, lack of, not

-haima blood

A pathological deficiency in the oxygen-carrying components of the blood.

Anemometer

Greek

anemos- wind

-meter (metron) instrument or means of measuring; to measure

Instrument used to measure wind speed.

Anesthesia

Greek

an- no, absence of, without, lack of, not

-aesthe- feeling, sensation, perception

-ia names of diseases, place names, or Latinizing plurals

Partial or total loss of the sense of pain, temperature, touch, etc., which may be produced by disease or an anesthetic.

Aneuploid

Greek

a- no, absence of, without, lack of, not

-neur- nerve

-nervus- sinew, tendon

-ploid having a number of chromosomes that has a specified relationship to the basic number of chromosomes

Aberration in the chromosome number, in which one or more extra chromosomes are present.

Aneurysm

Greek

an- no, absence of, without, lack of, not

-eurus- a widening; broad, wide

-ism state or condition, quality

Abnormal dilation of a blood vessel due to a congenital defect or weakness of the wall of the vessel.

Angialgia

Greek

angeion- vessel, usually a blood vessel

-algia pain, sense of pain; painful, hurting

Pain in a blood vessel.

Angiectasis

New Latin

angeion- vessel, usually a blood vessel

-ectasis expansion, dilation

Abnormal dilation of a blood vessel.

Angiitis

Greek

angeion- vessel, usually a blood vessel

-itis inflammation, burning sensation

Inflammation of a blood or lymph vessel.

Angina

Greek

ankhonē a strangling

A squeezing chest discomfort; angina pectoris occurs when blood oxygen is cut off from portions of the heart.

Angiocarditis

Greek

angeion- vessel, usually a blood vessel

-kard- heart, pertaining to the heart

-itis inflammation, burning sensation

Inflammation of the heart and great blood vessel.

Angiocarp

Greek

angeion- vessel, usually a blood vessel

-karpos fruit

A tree bearing fruit enclosed in a shell, involucrum, or husk.

Angiolith

Greek

angeion- vessel, usually a blood vessel

-lithe stone, rock

A calcareous deposit in the wall of a blood vessel.

Angiolysis

Greek

angeion- vessel, usually a blood vessel

-ly- (luein) to loosen, dissolve; dissolution, break

-sis action, process, state, condition

The obliteration of blood vessels, such as occurs during embryonic development.

Angionecrosis

Greek

angeion- vessel, usually a blood vessel

-nekros- death, corpse

-osis action, process, state, condition

Death of a blood vessel.

Angiosperm

Greek

angeion- vessel, usually a blood vessel

-sperma seed

Any of a class (Angiospermae) of vascular plants (such as orchids or roses) having the seeds in a closed ovary.

Angular

Latin

angulus angle

Having, forming, or consisting of an angle or angles.

Anhydride
Greek
an- no, absence of, without, lack of, not
-hydr- water
-ide binary compound
A chemical compound formed from another by the removal of water.

Anhydrous
Greek
an- no, absence of, without, lack of, not
-hydr- water
-ous full of, having the quality of, relating to
A compound in which all water has been removed, usually through heating.

Anisotropic
Greek
an- no, absence of, without, lack of, not
-isos- equal
-trope- bend, curve, turn, a turning; response to a stimulus
-ic (ikos) relating to or having some characteristic of
Not isotropic; having different properties in different directions; thus, crystals of the isometric system are optically isotropic, but all other crystals are anisotropic.

Annelid
Latin
annellus- little ring
-id state, condition; having, being, pertaining to, tending to, inclined to
Any of a phylum (Annelida) of coelomate and usually segmented invertebrates (such as earthworms, various marine worms, and leeches).

Anode
Greek
an- no, absence of, without, lack of, not
-hodós way or road
The negative terminal of a voltaic cell or battery.

Anomaly
Greek
an- no, absence of, without, lack of, not
-homolus- even
-y place for an activity, condition, state
The angular deviation, as observed from the sun, of a planet from its perihelion.

Anopheliphobia
Greek
an- no, absence of, without, lack of, not
-ophelos- advantage, use
-phob- fear, lacking an affinity for
-al of the kind of, pertaining to, having the form or character of
An abnormal fear or hatred of mosquitoes.

Anorexia
Greek
an- no, absence of, without, lack of, not
-orexis- appetite
-ia names of diseases, place names, or Latinizing plurals
Loss of appetite, sometimes because of a disease; anorexia nervosa.

Anoxia
Greek
an- no, absence of, without, lack of, not
-oxo- oxygen
-ia names of diseases, place names, or Latinizing plurals
Deprivation of oxygen that rapidly leads to collapse or death if not reversed.

Antacid
Greek
anti- opposing, opposite, against
-acere to be sour
Any substance that reduces stomach acid.

Antagonist
Greek
anti- opposing, opposite, against
-agon- conflict, contest
-ist one who is engaged in
A muscle or muscles that move in opposition to an agonist.

Antarctica
Greek
ante- before or prior to
-arc- bow arch or bent
-ic (ikos) relating to or having some characteristic of
A body of land found mostly south of the Artic Circle. It covers an area of 5,500,000 square miles. About 98% of the land mass is covered with a thick continental ice sheet, and the remaining 2% is barren rock.

Anterior
Latin
ante- before or prior to
-or a condition or property of things or persons
Located near or toward the head in lower animals.

Anther
Greek
anth- flower; that which buds or sprouts
-er one that performs an action
Pollen-bearing part of a stamen.

Antheridia
Greek/Latin
anth- flower; that which buds or sprouts
-oidium fungus

A sperm-producing organ occurring in seedless plants (fungi and algae).

Anthodite
Greek
anth- flower; that which buds or sprouts
-ite minerals and fossils
A period of the Paleozoic, spanning the time between 440 and 410 million years ago.

Anthophilous
Greek
anth- flower; that which buds or sprouts
-phile- one who loves or has a strong affinity or preference for
-ous full of, having the quality of, relating to
In biology, attracted to, or feeding on, flowers; living on or frequenting flowers.

Anthracite
Greek
anthrankitis- name of a fiery gem
-ite minerals and fossils
Hard coal that burns with very little smoke or flame.

Anthropic
Greek
anthropo- man; human being, mankind
-ic (ikos) relating to or having some characteristic of
Pertaining to humans or the period of their existence on earth.

Anthropobiology
Greek
anthropo- man; human being, mankind
-bios- life, living organisms or tissue
-logy (logos) used in the names of sciences or bodies of knowledge
The study of the biological relationships of humans as a species.

Anthropocentric
Greek
anthropo- man; human being, mankind
-kentron- center, sharp point
-ic (ikos) relating to or having some characteristic of
Regarding humans as the central element of the universe.

Anthropogenic
Greek
anthropo- man; human being, mankind
-gen- to give birth, kind, produce
-ic (ikos) relating to or having some characteristic of
Referring to pollutants and other impacts on natural environments that can be traced to human activities.

Anthropoid
Greek
anthropo- man; human being, mankind
-oid (oeidēs) resembling; having the appearance of
A group of primates that resemble humans; apes and monkeys.

Anthropology
Greek
anthropo- man; human being, mankind
-logy (logos) used in the names of sciences or bodies of knowledge
The scientific study of the history, culture, genetic conditions, and lifestyles of a given population of humans.

Anthropozoonosis
Greek
anthropo- man; human being, mankind
-zoon- animal
-nosis disease
An animal disease maintained in nature by animals and transmissible to humans.

Antibacterial
Greek
anti- opposing, opposite, against
-bacter- small rod
-ial (variation of *-ia*) relating to or characterized by
Pertaining to a substance that kills bacteria.

Antibiotic
Greek
anti- opposing, opposite, against
-bios- life, living organisms or tissue
-ic (ikos) relating to or having some characteristic of
Any of a large class of substances produced by various microorganisms having the power to arrest the growth of other microorganisms or to destroy them.

Antibody
Greek/Old English
anti- opposing, opposite, against
-botah (body) the material frame of humans and animals
Protein produced by the immune system in response to the presence of antigens in the body.

Anticline
Greek
anti- opposing, opposite, against
-klinein sloping, to lean
A fold of the rock strata that slopes downward from a center or common crest.

Anticoagulant
Latin
anti- opposing, opposite, against
-coāgulum- coagulator
-ant performing, promoting, or causing a specific event
A non–habit-forming medication that prevents the formation of clots in the blood.

Anticodon
Greek
anti- opposing, opposite, against
-caudex book
A sequence of three nucleotides found in t-RNA.

Anticyclone
Greek
anti- opposing, opposite, against
-kyklos- circle, wheel, cycle
-ne of or relating to
A system of winds rotating about a center of high atmospheric pressure, clockwise in the Northern Hemisphere and counterclockwise in the Southern, that usually advances at 20 to 30 miles (about 30 to 50 kilometers) per hour.

Antigen
Latin
anti- opposing, opposite, against
-gen to give birth, kind, produce
Substance to which the body responds by producing antibodies.

Antimatter
Greek
anti- opposing, opposite, against
-māter mother
A hypothetical form of matter that is identical to physical matter except that its atoms are composed of antielectrons, antiprotons, and antineutrons.

Antioxidant
Latin
anti- opposing, opposite, against
-oxy- pungent, sharp
-ant performing, promoting, or causing a specific event
A substance or enzyme that inhibits oxidation or inhibits the loss of an electron.

Antiparticle
Latin
anti- opposing, opposite, against
-particula a very small piece or part; a tiny portion or speck
A subatomic particle, such as a positron, antiproton, or antineutron, having the same mass, average lifetime, spin, magnitude of magnetic moment, and magnitude of electric charge as the particle to which it corresponds, but having the opposite sign of electric charge and opposite direction of magnetic moment.

Antisense
Greek/Latin
anti- opposing, opposite, against
-sentire to feel
Of or relating to a nucleotide sequence that is complementary to a sequence of messenger RNA. When antisense DNA or RNA is added to a cell, it binds to a specific messenger RNA molecule and inactivates it.

Antiseptic
Greek
anti- opposing, opposite, against
-sepsis- putrefaction or decay
-ic (ikos) relating to or having some characteristic of
Preventing or counteracting putrefaction or decay.

Antiserum
Greek/Latin
anti- opposing, opposite, against
-ser- the watery part of fluid
-um (**singular**) structure
-a (**plural**) structure
Animal or human serum containing antibodies that are specific to a number of antigens.

Antitoxin
Greek
anti- opposing, opposite, against
-toxikos- poison
-in protein or derived from protein
An antibody with the ability to neutralize a specific toxin.

Aortic
Latin
aort- lower extremity of the windpipe; by extension, extremity of the heart, the great artery
-ic relating to or having some characteristics of
Relating to the main trunk of the systemic arteries, carrying blood from the left side of the heart to the arteries of all limbs and organs except the lungs.

Apatite
Greek
apatē- deceit
-ite minerals and fossils
A natural, variously colored calcium fluoride phosphate, $Ca_5F(PO_4)_3$.

Aphasia
Greek
a- no, absence of, without, lack of, not
-phanai- speech

-ia names of diseases, place names, or Latinizing plurals

A condition characterized by defective or absent language abilities, typically caused by brain injury.

Aphelion
Greek
apo- away from
-helios- sun
-ion state, process, or quality of
The point on the orbit of a celestial body that is farthest from the sun.

Aphonia
Greek
a- no, absence of, without, lack of, not
-phonos- voice
-ia names of disease, place names, or Latinizing plurals
A condition characterized by the loss of one's voice, caused by a disease, injury to the vocal cords, or various psychological factors.

Aplasia
Greek
a- no, absence of, without, lack of, not
-plassein- to form
-ia names of disease, place names, or Latinizing plurals
Developmental failure of an organ or tissue to form, or the malformation of an organ or tissue.

Apnea
New Latin
a- no, absence of, without, lack of, not
-pnea breathing or breath
Temporary cessation of breathing.

Apocrine
Greek
apo- away from, off, separate
-krinein to separate
Applies to a type of mammalian sweat gland that produces a viscous secretion by breaking off a part of the cytoplasm of secreting cells.

Apoenzyme
Greek
apo- away from, off, separate
-en- in
-zuma leaven, yeast
The protein part of an enzyme to which the coenzyme attaches to form an active enzyme.

Apogee
Greek
apo- away from, off, separate
-gaia earth
Point of a satellite's orbit that is farthest from the sun.

Apogeotropism
Greek
apo- away from, off, separate
-geo- earth, world
-trope- bend, curve, turn, a turning; response to a stimulus
-ism state or condition
The response by an organism of turning away from the earth (e.g., plant stems growing upward).

Apomixis
Greek
apo- away from, off, separate
-mixis mingling, intercourse
Reproduction without meiosis, or the formation or fusion of gametes.

Aponeurosis
Greek
aponeurousthai to become tendinous
Sheetlike fibrous membrane that binds muscle to muscle or muscle to bone.

Apopyle
Greek
apo- away from, off, separate
-pyle gate
In sponges, opening of the radial canal into the spongocoel.

Apparatus
Latin
ad- to, a direction toward, addition to, near
-parare to make ready
A device or system composed of different parts that act together to perform some special function.

Appendage
Latin
ad- to, a direction toward, addition to, near
-pendere- to hang
-age (āticum) (**Latin**) condition or state
A part or an organ that is attached to the axis of the body (i.e., arm, leg); a structure arising from the surface or extending beyond the tip of another structure.

Appendectomy
Latin/Greek
ad- to, a direction toward, addition to, near
-pendere- to hang
(ectomy)
-ekt- outside, external, beyond
-tomos
(temnein) to cut, incise, section
The surgical removal of the vermiform appendix.

Appendicitis
Latin
ad- to, a direction toward, addition to, near
-pendere- to hang
-itis inflammation, burning sensation
An inflammation of the vermiform appendix.

Appendix
Latin
ad- to, a direction toward, addition to, near
-pendere to hang
A supplementary or accessory part of a bodily organ or structure.

Aquatic
Latin
aqua- water
-ic (ikos) relating to or having some characteristic of
Consisting of, relating to, or being in water; an organism that lives in, on, or near water.

Aquation
Latin
aqua- water
-ion state, process, or quality of
The process of replacement of other ligands by water.

Aqueous
Latin
aqua- water
-ous possessing, full of; characterized by
Relating to, similar to, containing, or dissolved in water.

Aquifer
Latin
aqua- water
-ferre to carry
Layer of rock or sediment that allows groundwater to pass freely.

Arachnid
Latin
arakhn- spider
-id state or condition; having, being, pertaining to, tending to, or inclined to
Arthropods characterized by four pairs of segmented legs and a body divided into two regions.

Arboraceous
French/Latin
erbe- herb
-aceous having the quality of
A reference to a tree or woodlike substance.

Arboreal
Latin
arbor- tree
-al of the kind of, pertaining to, having the form or character of
Of or pertaining to life in the trees or living things in the trees.

Archaeocytes
Greek
archae- original, beginning, origin, ancient
-cyte (kutos) sac or bladder that contains fluid
Amoeboid cells of varied functions in sponges.

Archaeology
Greek
archae- original, beginning, origin, ancient
-logy (logos) used in the names of sciences or bodies of knowledge
The study of past human life and culture by the recovery and examination of remaining material evidence.

Archaeopteryx
Greek
archae- original, beginning, origin, ancient
-pterux wing
A primitive group of birds existing in the Jurassic period, winged, with reptilian skin, teeth, and a long tail.

Archean
Greek
archae- original, beginning, origin, ancient
-an one that is of, relating to, or belonging to
The first formed rocks, characterized by cooling periods 3.8 to 2.5 billion years ago.

Archegonium
Greek
archae- original, beginning, origin, ancient
-gonos- offspring
-ium quality or relationship
A flasklike reproductive organ found in mosses, ferns, and some other gymnosperms where the eggs are produced.

Archenteron
Greek
archae- original, beginning, origin, ancient
-enteron gut
The main cavity of an embryo in the gastrula stage.

Archeognatha
Greek
archae- original, beginning, origin, ancient
-gnatha jaw
Bristletail; insect with cylindrical body, no wings, and three terminal "tails" with a medial caudal filament. Found in rocky areas, it is crepuscular or nocturnal.

Archetype
Greek
archae- original, beginning, origin, ancient
-tupos type, model, stamp
An original model or pattern from which copies are made or evolve.

Area
Latin
area open space
The extent of a planar region or of the surface of a solid measured in square units.

Areola
Latin
area- a courtyard, open space
-ola little
A small ring of color around a center portion, as about the nipple of the breast, or the part of the iris surrounding the pupil of the eye.

Argillaceous
Latin
argillos- clay
-aceous having the quality of
Of the nature of clay; largely composed of clay.

Argon
Greek
a- no, absence of, without, lack of, not
-ergon work
A colorless, inert gaseous element composing approximately 1% of the earth's atmosphere.

Arillate
Latin
arillus- grape seed
-ate characterized by having
A seed with an unusually brightly colored cover.

Arithmetic
Greek
arithmos- number
-ic (ikos) relating to or having some characteristic of
The computation of numbers having to do with addition, subtraction, multiplication, and division.

Aromatic
Greek
aroma- smell (due to sweet smell of benzene and related organic groups)
-ic (ikos) relating to or having some characteristic of
Of, relating to, or containing one or more six-carbon rings characteristic of benzene series and related organic groups.

Arteriole
Greek
arteria windpipe, artery
-ole little
Small, terminal branch of an artery that leads into a capillary bed.

Arteriomalacia
Greek
arteria- windpipe, artery
-malacia softening of tissue
The softening of arteries, usually as a result of some disorder.

Arteriosclerosis
Greek
arteria- windpipe, artery
-sklero- (sklēroun) to harden
-sis action, process, state, condition
A chronic disease in which thickening, hardening, and loss of elasticity of the arterial walls result in impaired blood circulation.

Artery
Greek
arteria windpipe, artery
A vessel that carries blood from the heart to the cells, tissues, and organs of the body.

Arthralgia
Greek
arthr- joint
-algia pain, sense of pain; painful, hurting
Pain resulting from inflammation in a joint.

Arthritis
Greek
arthr- joint
-itis inflammation, burning sensation
An inflammation of a joint.

Arthroplasty
Greek
arthr- joint
-plastos- (plassein) something molded (to mold)
-y place for an activity; condition, state
Surgical reconstruction or replacement of a malformed or degenerated joint.

Arthropod
Greek
arthr- joint
-poda foot
Any of numerous invertebrate animals of the phylum Arthropoda, including insects, crustaceans, arachnids, and myriapods.

Arthroscopy
Greek
arthr- joint
-skopion for viewing with the eye
Visual examination of the inside of a joint with the use of a specialized scope.

Astrology

The ancient Greeks bore witness to the orderly nature of the daytime and nighttime skies. Based on this recognition, they gave the name *cosmos*, meaning "order," to the celestial sphere. The serenity of the cosmos apparently gave the ancients a sense of security from the knowledge that tomorrow's nighttime sky would closely resemble tonight's.

The Mesopotamians are credited with the advent of Western astrology in the second millennium BC. They believed that the arrangement of the stars and planets somehow influences human existence here on earth. The term *zodiac* was given to an imaginary band or belt spanning about 8 degrees on either side of the path of the sun. *Zodiac* comes from the Greek word *zoon*, meaning "animal" or "animal-like," reflecting the fact that the major constellations in the band are named after animals or animal-like creatures. The pathway defined by the zodiac also includes the orbital paths of many planets in our solar system as well as our moon. The Greeks are credited with the cre-

ation of the horoscope, which is a chart prepared at the conception of a particular human being. By plotting stellar and planetary positions in the zodiac, ancient astrologers believed that the course of one's life could be foretold. So skillful were these Greeks in the use of astrological charts and prediction that over the course of human history few changes have been made to the methodology of astrology as practiced by the Greeks.

Astrology, of course, is a pseudoscience. However, among the early Arab astrologers and later in both Jewish and Christian sects, astrology developed into a vital component of the relationship between man and his deity.

Astrology is as popular among the public today as it was during the Middle Ages and before, especially in the United States. Scientists discount any relationship between the positions of heavenly bodies and prognosticative power. Most treat astrology as it should be treated, as a source of amusement and fun.

Articulation

Latin
articulus- small joint
-ate- of or having to do with
-ion state, process, or quality of
The action of bending the joints; a movable or fixed joint between two or more bones.

Artificial

Latin
artificialis- not natural, man-made
-ial relating to or characterized by
Produced by humans rather than occurring naturally; refers to something created or modified through the effects of human or sociological forces.

Artiodactyla

Greek
artios- even
-daktulos toe, finger, digit
Order including even-toed mammals (deer, cows, sheep).

Asbestos

Greek
a- no, absence of, without, lack of, not
-sbennunai to quench
Magnesium silicate; a fibrous, incombustible, and chemical resistance substance used for fireproofing and insulation.

Ascarid

Greek
askarizein- to jump, throb
-id state or condition; having, being, pertaining to, tending to, or inclined to
Any of a family of nematode worms, including the common roundworm *(Ascaris lumbricoides)*, which is parasitic in the human intestine.

Ascocarp

Greek
askos- bag
-karpos fruit
The mature, saclike fruiting body of an ascomycetes fungi.

Ascomycetes

Greek
askos- bag
-mukēs fungus
A class of fungi containing an ascus and spores.

Ascus

Greek
askos- bag
A saclike spore capsule located at the tip of the ascocarp in the phylum Ascomycota.

Asepsis

Latin

a- no, absence of, without, lack of, not
-sepein- to decay, cause to rot
-sis action, process, state, or condition
The absence of contamination by unwanted organisms.

Aseptic
Greek
a- no, absence of, without, lack of, not
-sepein- to decay, cause to rot
-ic (ikos) relating to or having some characteristic of
Pertaining to the condition of being free from germs or other infection-causing microorganisms.

Asexual
Latin
a- no, absence of, without, lack of, not
-sexus sex
Refers to reproduction in which a single parent produces offspring that are genetically identical to the parent.

Asphyxia
Greek
a- no, absence of, without, lack of, not
-sphyzein- to throb; pulse, heartbeat
-ia names of diseases, place names, or Latinizing plurals
A condition in which an extreme decrease in oxygen in the body accompanied by an increase in the concentration of carbon dioxide leads to loss of consciousness or death.

Aspiration
Latin
a- no, absence of, without, lack of, not
-spir- breath of life, breath, breathing
-ion state, process, or quality of
The process of withdrawing fluid from a cavity or sac by the use of a needle.

Assay
Latin
assa- pure, whole
-y place for an activity; condition or state
In chemistry, the determination of the quality of a substance present in a sample.

Assimilate
Latin
ad- to, a direction toward, addition to, near
-simulare- to make similar or alike
-ate characterized by having
To consume, digest, absorb, and assimilate nutrients into a living being.

Assimilation
Greek

ad- to, a direction toward, addition to, near
-simulare- to make similar or alike
-ion state, process, or quality of
Process by which absorbed food molecules circulating in the blood pass into the cells and are used for growth, tissue repair, or other metabolic activities.

Astatine
Greek
a- no, absence of, without, lack of, not
-statos- standing, stay, make firm, fixed, balanced
-ine in a chemical substance
A highly unstable, radioactive element.

Asteroid
Greek
aster- star
-oid (oeidēs) resembling; having the appearance of
Any of the small celestial bodies between the orbits of Mars and Jupiter.

Asteroidea
Greek
aster- star
-oid (oeidēs) resembling; having the appearance of
Any of various marine echinoderms of the class Asteroidea, characteristically having a thick, often spiny body with five arms extending from a central disk.

Asthenia
Greek
a- no, absence of, without, lack of, not
-sthenos- strength
-ia names of diseases, place names, or Latinizing plurals
Loss or lack of bodily strength or energy; weakness, debility.

Asthenosphere
Greek
a- no, absence of, without, lack of, not
-sthenos- strength
-sphaira a globe shape, ball, sphere
A layer of hot, weak material located in the mantle at a depth between 100 and 350 kilometers; the rock within the zone is easily deformed.

Astigmatism
Greek
a- no, absence of, without, lack of, not
-stigma- a point, mark, spot, puncture
-ism state or condition, quality
A defect in an optical system (i.e., impaired eyesight) in which light rays fail to converge to a single focal point.

Galileo (1564–1642)

Galileo Galilei was born on February 15, 1564, in the Tuscan region of Italy. His accomplishments in the sciences are far too extensive to be covered in a brief exposé. He spent most of his life studying mathematics, astronomy, and physics. He was a Catholic and had many friends who held esteemed positions in the Catholic Church, but he found himself on the defensive for his support of the heliocentric configuration of the solar system as described by Copernicus. For this position, in his later years, he was put on trial and confined to house arrest for the remaining days of his life.

Galileo is given credit for inventing the telescope; he actually did not invent it but rather refined and improved its design. With the advent of the lens, he created a telescope that enabled him to observe and study sunspots. This probably contributed to his loss of sight. He made it possible to see, for the first time, the moons orbiting Jupiter. His observations of Venus and its phases, which were much like the phases of the moon, led Galileo to side with the Copernican, heliocentric model of the solar system rather than the widely accepted geocentric model put forth by Ptolemy. Galileo sold quite a few of his telescopes and made a handsome profit marketing them to seafarers.

Galileo is hailed as the standard-bearer for scientific methodology. Influenced by his strong background in mathematics, he advocated and pioneered experimental designs that included quantification of data. This was a dramatic departure from earlier practices in science, where a more philosophical, qualitative approach was the norm. For this and other reasons, Galileo stood at odds with the Church and with the more traditional, Aristotelian thinkers. Looking back at his rather radical departure from older approaches to science, we acknowledge Galileo as the father of science. He is also credited as the father of modern physics and of modern astronomy.

We can confirm that Galileo had more than a casual interest in technology. He developed a thermometer using an enclosed tube, water, and objects floating in the water. It operates on the principles of temperature, compressed air and buoyancy, and displacement. He designed and developed the first compound microscope with concave and convex lenses. Galileo also created a vastly improved version of the military compass, paving the way for improved weaponry. His military compass provided a much safer way of elevating and supporting cannons, increasing their firepower and accuracy.

Galileo studied pendulums and noted that the period of the swing is independent of the wave's amplitude. The advent of the pendular clock later developed by Christian Huygens depended on the development of the escapement mechanism for the pendulum created by Galileo.

His work in physics is well known and continues to be discussed in schools today. Recall his experiment with two balls of unequal mass dropped from the Tower of Pisa. He contended that the time of descent of a ball was independent of its mass. This was the exact opposite of what Aristotle had proposed centuries before. Even though Galileo was not the first person to make this argument, he was able to demonstrate using inclined planes and rolling balls that the principle was indeed correct.

Astrobiology

Greek

astros- star

-bios- life, living organisms or tissue

-logy (logos) used in the names of sciences or bodies of knowledge

The branch of biology that deals with the search for extraterrestrial life and the effects of extraterrestrial surroundings on living organisms.

Astrocyte

Greek

astros- star

-cyte (kutos) sac or bladder that contains fluid

A star-shaped cell, especially a neuroglial cell of nervous tissue.

Astrology

Greek

astros- star

-logy (logos) used in the names of sciences or bodies of knowledge

The study of the positions of the stars and planets based on the belief that they can predict the future.

Astronaut

Greek

astros- star

-nautes sailor

A traveler in space; a member of a U.S. space crew trained to pilot, navigate, or conduct research in outer space.

Astronomy
Greek
astros- star
-nom (nemein) to dictate the laws of; knowledge, usage, order
Study of planets, stars, and other objects in space.

Astrophysics
Greek
astros- star
-phusis- nature
-ic (ikos) relating to or having some characteristic of
The branch of astronomy that deals with the physics of stellar phenomena.

Asymmetric
Greek
a- no, absence of, without, lack of, not
-summetros- of like measure
-ic (ikos) relating to or having some characteristic of
Unequal in size or shape; having no balance.

Asymptotic
Greek
a- no, absence of, without, lack of, not
-sumptotos intersecting
Refers to a line whose distance to a given curve tends to zero; an asymptote may or may not intersect its associated curve.

Asystole
Greek
a- no, absence of, without, lack of, not
-sustellein to contract
A life-threatening cardiac condition marked by failure of the heart to contract.

Atactic
Greek
a- no, absence of, without, lack of, not
-taktos ordered
The type of orientation of the methyl groups on a polypropylene chain in plastics—in this case random orientation.

Ataxia
Greek
a- no, absence of, without, lack of, not
-taxis order
Loss of the ability to coordinate muscular movements.

Athermancy
Greek
a- no, absence of, without, lack of, not
-thermos- combining form of "hot" (heat)
-ancy condition or state of
Impermeability to heat (i.e., no heat passing through); the inability to transfer radiant energy.

Athermy
Greek
a- no, absence of, without, lack of, not
-thermos- combining form of "hot" (heat)
-y place for an activity; condition or state
A therapeutic treatment for certain diseases involving no heat.

Atherosclerosis
Greek
athera- tumors full of pus, like a gruel
-skleros- hardening
-sis action, process, state, condition
A stage of arteriosclerosis involving fatty deposits (atheromas) inside the arterial walls.

Atmosphere
Greek
atmos- vapor
-sphaira a globe shape, ball, sphere
Mixture of gases that surrounds the earth.

Atoll
Sanskrit
antara interior
A nearly circular coral reef surrounding a shallow lagoon.

Atom
Greek
a- no, absence of, without, lack of, not
-tomos (temnein) to cut, incise, section
A unit of matter, the smallest of an element, having all the characteristics of that element and consisting of a dense, positively charged nucleus surrounded by an electron cloud.

Atonia
Greek
a- no, absence of, without, lack of, not
-tonos- tone, stretching, firm
-ia names of diseases, place names, or Latinizing plurals
Decrease in or lack of normal muscle tone, sometimes caused by prolonged paralysis.

Atrioventricular
atri- open area, central court, hall, entrance, or main room of an ancient roman house
-ventricul- belly
-ar relating to or resembling
Relating to, involving, or resembling the area of the atrium or ventricle of the heart; the atrioventricular valve.

Atrium
Latin
atri- open area, central court, hall, entrance, or main room of an ancient roman house

-ium quality or relationship
Chamber associated with the heart; upper chamber.

Atrophy
Greek
a- no, absence of, without, lack of, not
-trophos- (trophein) to nourish, food, nutrition; development
-y place for an activity; condition, state
A wasting away, deterioration, diminution, or decrease in the size of a body organ, tissue, or part owing to disease, injury, or lack of use.

Attenuate
Latin
ad- to, a direction toward, addition to, near
-tenuis- thin
-ate of or having to do with
To make or become weaker; to reduce the size, strength, or density of something; to become thinner, weaker, less dense, or less virulent.

Auditory
Latin
audit- hearing, listening, perception of sounds
-ory tending to, serving for
Of or relating to hearing, the organs of hearing, or the sense of hearing.

Auricle
Latin
auricula ear
An ear-shaped part of an organ.

Aurora
Latin
aurora dawn
Short for *aurora australis or aurora borealis* (luminous bands or streamers of light visible in night sky).

Aurous
Latin
aurum- gold
-ous full of, having the quality of, relating to
Of, relating to, or containing gold.

Austral
Latin
austr- south; south wind
-al of the kind of, pertaining to, having the form or character of
Relating to or coming from the south.

Australopithecus
Latin
austral- southern; human race classification
-pithecus ape, apelike creatures
Extinct genus of African hominid family thought to have lived between 4 and 1 million years ago.

Autecology
Greek
auto- self, same, spontaneous; directed from within
-oikos- home, house
-logy (logos) used in the names of sciences or bodies of knowledge
The ecology of an individual organism or species.

Autism
Greek
auto- self, same, spontaneous; directed from within
-ism state or condition, quality
A psychiatric disorder of childhood characterized by marked deficits in communication and social interaction, preoc-cupation with fantasy, language impairment, and abnormal behavior, such as repetitive acts and excessive attachment to certain objects.

Autoclave
French
auto- self, same, spontaneous; directed from within
-clavis key (from the fact that it's self-locking from the pressurization)
A strong, pressurized, steam heat vessel, as used for laboratory experiments, sterilization, or cooking.

Autogenous
Greek
auto- self, same, spontaneous; directed from within
-gen- to give birth, kind, produce
-ic (ikos) relating to or having some characteristic of
Self-generated; produced independently. Coming from the individual that it is growing in; a graft.

Autoionization
Greek
auto- self, same, spontaneous; directed from within
-ion- (ienai) to go; something that goes
-izein to cause or become
-ion state, process, or quality of
An ionization reaction between identical molecules.

Autolysis
Greek
auto- self, same, spontaneous; directed from within
-ly- (luein) to loosen, dissolve; dissolution, break
-sis action, process, state, condition
Self-acting disintegration of tissue by the release of enzymes within the cells.

Autonomic
Greek
auto- self, same, spontaneous; directed from within
-nom (nemein) to dictate the laws of; knowledge, usage, order
-ic (ikos) relating to or having some characteristic of
Functioning independently of the will; not under voluntary control (e.g., as with most functions of the nervous system).

Autopsy
Greek
auto- self, same, spontaneous; directed from within
-opsy examination
Examination of the organs of a body to determine the cause of death.

Autosomal
Greek
auto- self, same, spontaneous; directed from within
-soma (somatiko) body
-al of the kind of, pertaining to, having the form or character of
Pertaining to or characteristic of an autosome.

Autosome
Greek
auto- self, same, spontaneous; directed from within
-soma (somatiko) body
Any chromosome other than those that determine the sex of an organism.

Autotherm
Greek
auto- self, same, spontaneous; directed from within
-thermos combining form of "hot" (heat)
An organism that regulates its body heat independently of ambient temperature changes.

Autotoxin
Greek
auto- self, same, spontaneous; directed from within
-toxikos poison
Any harmful substance generated within the body; something that is self-poisonous.

Autotroph
Greek
auto- self, same, spontaneous; directed from within
-trophos (trophein) to nourish; food, nutrition; development
An organism that makes organic nutrients from inorganic raw materials; any organism considered to be a producer, capable of making its own food.

Autotrophic
Greek
auto- self, same, spontaneous; directed from within
-trophos- (trophein) to nourish; food, nutrition; development
-ic (ikos) relating to or having some characteristic of
Relating to the process of synthesizing food either by photosynthesis or by chemosynthesis.

Auxin
Greek
auxein to grow
Any of several plant hormones that regulate various functions, including cell elongation.

Average
Arabic
awariyah damaged merchandise
A single value that summarizes or represents the general significance of a set of unequal values.

Avian
Latin
avis bird
Of, relating to, or characteristic of birds.

Aviation
Latin
avis- bird
-ation state, process, or quality of
The art or science of flying, especially airplanes.

Avicide
Latin
avis- bird
-cide (caedere) to cut, kill, hack at, or strike
Type of pesticide that controls populations of birds considered to be pests.

Axiom
Greek
axios worthy
A universally recognized truth; self-evident, established rule.

Axis
Latin
axis central
Any of the anatomical structures that lie centrally or along a midcentral line within a body.

Axon
Greek
axōn axis
The usually long process of a nerve fiber that generally conducts impulses away from the body of the nerve cell.

Azeotrope
Greek
a- no, absence of, without, lack of, not
-zein- to boil
-trope bend, curve, turn, a turning; response to a stimulus
A mixture of two or more substances that has the same composition in vapor state and liquid state.

Azimuth
Arabic
al- the
-samt way, path
In astronomy, the horizontal measurement of the position of an object from north to east (clockwise) in degrees from a reference direction or a celestial body (polaris).

B

Bacteremia
Greek
baktron- staff, rod
-haima- blood
-ia names of diseases, place names, or Latinizing plurals
Presence of bacteria in the blood.

Bacteria
Greek
baktron- staff, rod
-ia names of diseases, place names, or Latinizing plurals
Single-celled or noncellular spherical or spiral- or rod-shaped organism without chlorophyll.

Bactericide
Latin
baktron- staff, rod
-cida cutter, killer, slayer.
Any chemical agent that kills bacteria

Bacteriophage
Greek
baktron- staff, rod
-phagein to eat
An ultra-microscopic filter-passing agent that has the power to destroy bacteria and to induce bacterial mutation.

Bacteriostat
Greek
baktron- staff, rod
-statos standing; stay; make firm, fixed, balanced
A class of antibiotics that prevents growth of bacterial cells.

Bacteriotherapy
Greek
baktron- staff, rod
-therapeuein heal, cure; treatment
Treatment of disease by introducing bacteria into the system.

Bacteriotropic
Greek
baktron- staff, rod
-trope- bend, curve, turn, a turning; response to a stimulus
-ic (ikos) relating to or having some characteristic of
Having an affinity for bacteria; moving toward bacteria.

Bacterium
Greek
baktron- staff, rod
-ium quality or relationship
A single-celled or non-cellular spherical or spiral- or rod-shaped organism lacking chlorophyll that reproduces by fission; important as a pathogen and for its biochemical properties; taxonomy is difficult (often considered a plant).

Bacteroid
Greek
baktron- staff, rod
-oid (oeidēs) resembling, having the appearance of
Resembling bacteria in appearance or action.

Barometer
Greek
baro- weight, heavy; combining form meaning "pressure"

-meter (metron) instrument or means of measuring; to measure
An instrument for determining the weight or pressure of the atmosphere, and hence used for judging probable changes in the weather.

Baroreceptor
Greek
baro- weight, heavy; combining form meaning "pressure"
-reciepere- to receive
-or a condition or property of things or persons; person that does something
In living tissue, a receptor end organ that responds to pressure.

Base
Latin
basis fundamental ingredient, foundation
Any large class of compounds, including the hydroxides and oxides of metals, having the ability to react with acids to form salts.

Basidiomycete
Latin/Greek
basid- foundation or base
-idion- (**Greek**) diminutive suffix
-mukēt fungus
Any of a large group of fungi, including puffballs, shelf fungi, rusts, smuts, and mushrooms, that bear sexually produced spores on a basidium.

Basidium
Latin
basid- foundation or base
-ium quality or relationship
Club-shaped organ involved in sexual reproduction in basidiomycete fungi (mushrooms, toadstools etc.) and bearing four haploid basidiospores at its tip.

Basophile
Greek
basis- fundamental ingredient, foundation
-phile one who loves or has a strong affinity or preference for
A granulocytic white blood cell characterized by cytoplasmic granules that stain blue when exposed to a basic dye.

Batholith
Greek
bathy- deep, depth
-lith rock, stone
A mass of igneous rock that has melted and intruded into surrounding strata.

The Greek Language

Examining the origins of the languages of Western cultures, we see that most had their beginnings in the language of the Greeks. Around the sixth century BC, the ancient Greek culture flourished. Democracy, cherished only by the wealthy, provided a political and social environment for philosophers to ponder the nature of the universe. Some put down in words their interpretations of order and chaos. Plato (427–347 BC), one of the most famous Greek philosophers, metaphorically linked science to politics by stating that all things celestial were pure and godly while earthly things were somehow tarnished and corrupted. He referred to planets as crystalline spheres and made an analogy between the good and the sun: "though the good itself is not essence but still transcends essence in dignity and surpassing power." In Plato's *Allegory of the Cave* he speaks of shadows and captivity and the darkness. In many such ways Plato and others advanced the sciences in their time. Yet some would say they also suppressed science and philosophy through their belief that these endeavors befit only the elite in Greek society.

Bathyal
Greek
bathy- deep, depth
-al of the kind of, pertaining to, having the form or character of
Of or relating to a region of the ocean between depths of 200 and 4,000 meters (660 and 13,000 feet).

Bedrock
Old English/Latin
bed- bed
-rocca rock, stone
The layer of solid rock beneath the gravel, soil, and stone of the earth's surface.

Behavior
Old English/French
be- to cause, make, affect
-havour to have
In biology, all of the responses to stimuli that an organism is capable of displaying.

Benthic
Greek
benthos- bottom
-ic (ikos) relating to or having some characteristic of
Of the benthos, or bottom of the ocean or deep lake; organisms existing at the bottom zone of the sea.

Beta (rays)
Greek
beta second letter of the Greek alphabet
Electrons or positrons that are emitted from a radioactive substance.

Bias
French
biais slant
To apply a small voltage to.

Bicephalous
Greek
bi- two, twice, double, twofold
-cephalo- (kephalikos) head
-ous full of, having the quality of, relating to
Having two heads.

Bicuspid
Latin
bi- two, twice, double, twofold
-cuspis- sharp point, cusp
-id state, condition; having, being, pertaining to, tending to, inclined to
Having two points or cusps, such as a premolar tooth.

Bidentate
Greek
bi- two, twice, double, twofold
-dentis- tooth
-ate to cause to be affected or modified by
To have two teeth or teethlike parts.

Bifurcation
Latin
bi- two, twice, double, twofold
-furca- fork
-ation state, process, or quality of
The point at which a splitting into two pieces occurs.

Bilateral
Latin
bi- two, twice, double, twofold
-latus- side
-al of the kind of, pertaining to, having the form or character of
Referring to two-sided symmetrical animals; having identical parts on each side of an axis.

Bilirubin
Latin

bilis- bile
-ruber- red
-in protein or derived from protein
A pigmented substance in the hemoglobin that appears in the urine, darkening it and indicative of liver or gallbladder disease.

Bimetallic
Latin
bi- two, twice, double, twofold
-metallon- mine, ore, quarry; any of a category of electropositive elements from metallum
-ic (ikos) relating to or having some characteristic of
Relating to a substance composed of two different metals that are bonded together.

Binary
Latin
bini- two at a time, two by two
-ary of, relating to, or connected with
Consisting of or involving two, as in binary fission.

Binocular
Latin
bi- two, twice, double, twofold
-ocul- of or relating to the eye
-ar relating to or resembling
Having two eyes arranged to produce stereoscopic vision.

Binomial
Latin
bi- two, twice, double, twofold
-nom- (nemein) to dictate the laws of; knowledge, usage, order
-al of the kind of, pertaining to, having the form or character of
A taxonomic name consisting of two terms; binomial nomenclature.

Bioaccumulation
Greek/Latin
bios- life, living organisms or tissue
-ad- to, a direction toward, addition to, near
-cumulāre- to pile up
-ion state, process, or quality of
To accumulate in a biological system.

Bioaugmentation
Greek/Latin
bios- life, living organisms or tissue
-augere- to increase
-ion state, process, or quality of
Increasing the activity of bacteria that decompose pollutants, a technique used in bioremediation.

Biocentrism
Greek
bios- life, living organisms or tissue

-kentron- center, sharp point
-ism state or condition, quality
The belief that all life—or even the whole universe, living or otherwise—taken as a whole, is equally valuable, and that humanity is not the center of existence.

Biodegradable
Greek
bios- life, living organisms or tissue
-degrade- to impair physical structure
-able capable, inclined to, tending to, given to
Capable of being decomposed by biological agents, especially bacteria.

Biodiversity
Greek
bios- life, living organisms or tissue
-diverse- differing from another
-ity state, quality
The number and variety of organisms found within a specified region.

Bioecologist
Greek
bios- life, living organisms or tissue
-eco- environment, habitat
-logist a person who studies
A specialist who studies the relation-ships of organisms to their natural environments.

Bioenrichment
Greek/Latin/French
bios- **(Greek)** life, living
en- **(Latin)** in
-riche- **(French)** rich
-ment state or condition resulting from a (specified) action
Adding nutrients or oxygen to increase the microbial breakdown of pollutants.

Biofuel
Various
bios- life, living organisms or tissue
-focus (fuel) hearth, fireplace
Any fuel derived from biomass, such as treated municipal and industrial wastes and methane produced from renewable resources, especially plants.

Biogenesis
Greek
bios- life, living organisms or tissue
-gen- to give birth, kind, produce
-sis action, process, state, condition
The biological principle that life originates or arises from life, and not from nonliving things.

Biogeography
Greek

bios- life, living organisms or tissue
-geo- earth
-graphia (graphein) to write, record, draw, describe
The study of the geographical distribution of organisms.

Biolith
Greek
bios- life, living organisms or tissue
-lithos stone or rock
A rock of organic origin.

Biologics
Greek
bios- life, living organisms or tissue
-logics talk, speak; speech; word
Agents, such as vaccines, that confer immunity to diseases or harmful biotic stresses.

Biology
Greek
bios- life, living organisms or tissue
-logy (logos) used in the names of sciences or bodies of knowledge
The science of life and of living organisms, including their structure, function, growth, origin, evolution, and distribution.

Biomass
Greek
bios- life, living organisms or tissue
-maza mass, large amount
The total amount or weight of living material in a given area.

Biome
Greek
bios- life, living organisms or tissue
-oma community
A major region, such as continental grassland, that has similar physical and climatological conditions.

Biomimesis
Greek
bios- life, living organisms or tissue
-minie- mimic, mime; imitate, act; simulation
-sis action, process, state, condition
In biology, the ability of an organism to mimic the physical characteristics of another species.

Biomimetics
Greek
bios- life, living organisms or tissue
-minie- mimic, mime; imitate, act; simulation
-ic (ikos) relating to or having some characteristic of
A branch of biology that uses information from biological systems to develop synthetic systems.

Biopesticide

Latin/Greek
bios- life, living organisms or tissue
-pestis- (**Latin**) plague, pestilance
-cide (caedere) to cut, kill, hack at, or strike
Naturally occurring substances with pesticidal properties.

Biopsy

Greek
bios- life, living organisms or tissue
-opsy examination
Selection of tissue removed from a living specimen.

Bioremediation

Greek
bios- life, living organisms or tissue
-re- again
-medi- middle
-ion state, process, or quality of
The process of using bacteria or other organisms to "clean up" toxins in the environment.

Biosphere

Greek
bios- life, living organisms or tissue
-sphaire to surround
The thin outer shell of the earth and the inner layers of its atmosphere, the place where all living systems are found.

Biotechnology

Greek
bios- life, living organisms or tissue
-tekhne- skill, systematic treatment
-logy (logos) used in the names of sciences or bodies of knowledge
The scientific manipulation of living organisms, especially at the molecular genetic level, to produce useful products. Gene splicing and the use of recombinant DNA(rDNA) are major techniques used.

Biotic

Greek
bios- life, living organisms or tissue
-ic (ikos) relating to or having some characteristic of
Living materials in an ecosystem; having some characteristics of living organisms.

Biotoxin

Greek
bios- life, living organisms or tissue
-toxikos poison
Any toxic substance formed in an animal body and demonstrable in its tissues or body fluids, or both.

Bipectinate

Latin
bi- two, twice, double, twofold
-pectin- comb
-ate characterized by having
Feathery, with comblike branches or projections growing out from both sides of the main axis (applied mainly to insect antennae).

Bipedal

Latin
bi- two, twice, double, twofold
-ped- foot
-al of the kind of, pertaining to, having the form or character of
An organism having two feet or capable of walking on two feet.

Biramous

Latin
bi- two, twice, double, twofold
-ramus- branch
-ous full of, having the quality of, relating to
Consisting of or having two branches, as the appendages of an arthropod.

Bitumen

Latin
bitūmen a mineral pitch from the Near East
Any of various flammable mixtures of hydrocarbons and other substances, occurring naturally or obtained by distillation from coal or petroleum, that are components of asphalt and tar and are used for surfacing roads and for waterproofing.

Bivalve

Latin
bi- two, twice, double, twofold
-valve leaf of a door
A mollusk that has a shell consisting of two hinged valves.

Bladder

Latin
blaedre bladder
In biology, any sac or saclike organ that is capable of distension as it fills with fluid.

Blastocoel

Greek
blastos- germ, bud
-koilos hollow
Cavity of the blastula.

Blastocyst

Greek
blastos- germ, bud
-kustis (cyst) sac or bladder that contains fluid
The modified blastula that is characteristic of placental mammals.

Blastomere
Greek
blastos- germ, bud
-meros part
Name given to the early group of cells that result from the fertilization and cleavage of an ovum.

Blastopore
Greek
blastos- germ, bud
-poros passage
The opening of the archenteron (the central opening of the gastrula, which ultimately becomes the digestive cavity).

Blastula
Greek
blastos- germ, bud
-ula diminutive
Early embryological stage of many animals; consisting of a hollow mass of cells.

Blennogenic
Greek
blenno- mucus
-gen- to give birth, kind, produce
-ic relating to or having some characteristic of
Producing or secreting mucus.

Blepharoplast
Greek
blepharon- eyelid
-plastos (plassein) something molded; to mold
A very small mass of cytoplasm at the base of a flagellum, containing small amounts of chromatin.

Blood
Old English
blōd to thrive or bloom
The fluid consisting of plasma, cells, and platelets that is circulated by the heart through the vertebrate vascular system.

Bomb
Greek
bombos booming sound
A container capable of withstanding high internal pressure.

Boreal
Latin
boreios coming from the north
Northern; of or relating to the north; the north wind.

Botany
Greek
botanē- fodder, plants
-onuma name
The science or study of plants.

Botulism
Latin
botulus- sausage
-ism state or condition, quality
A severe, sometimes fatal poisoning caused by ingestion of food containing botulin and characterized by nausea, vomiting, disturbed vision, muscular weakness, and fatigue.

Boule
Latin
bulla bubble
A pear-shaped, aluminum-based synthetic mineral.

Bovine
Latin
bov- cow
-ine of or relating to
Relating to, affecting, resembling, or derived from a cow or bull.

Bowel
Latin
botulus sausage
The intestines; sometimes refers to the large intestine.

Brachial
Greek
brackhiōn upper arm
-al of the kind of, pertaining to, having the form or character of
Of or relating to the arm, forelimb, or wing of a vertebrate.

Brachiopod
Greek
brakhin- upper arm
-pod foot
Any of various marine invertebrates of the phylum Brachiopoda, having bivalve dorsal and ventral shells enclosing a pair of tentacled, armlike structures that are used to sweep minute food particles into the mouth; also called lampshell.

Brachiosaurus
Greek
brakhin- upper arm
-sauros lizard
The group of very large, herbivorous dinosaurs existing in the Jurassic and Cretaceous periods; notable features include long forelegs and a long neck.

Bradycardia
Latin/Greek
bradus- slow
-kard- heart; pertaining to the heart
-ia names of diseases, place names, or Latinizing plurals

Slower-than-normal heart rate in humans, usually considered to be less than 60 beats per minute.

Breeds
Old English
bredan to breed
Variations within the same species that are capable of reproducing with one another; phenotypic modifications within a group.

Brevis
Latin
brevis brief
An anatomical term meaning "short," usually associated with skeletal muscle.

Brittle
Old English
brytel to shatter
Likely to break, snap, or crack.

Bronchitis
Greek
bronkhos- windpipe
-itis inflammation, burning sensation
Chronic or acute inflammation of the mucous membrane of the bronchial tubes.

Bronchogenic
Greek
bronkhos- windpipe
-gen- to give birth, kind, produce
-ic (ikos) relating to or having some characteristic of
Originating in the bronchi or having its origin in the bronchus.

Bronchomalacia
Greek
bronkhos- windpipe
-malacia softening of tissue
The degeneration or softening of the trachea as a result of some disorder.

Bronchus
Greek
bronkhos- windpipe
-us singular
Main branch of the windpipe.

Bryophyte
Greek
bruein- to swell or teem
-phyte plant
Any of a division of nonvascular plants that lack vascular tissue, including mosses and liverworts.

Bryozoan
Greek
bruon- moss
-zôion living being
Any of various small aquatic animals of the phylum Bryozoa that reproduce by budding and form mosslike or branching colonies permanently attached to stones or seaweed; also called moss animal or polyzoan.

Buoyancy
Dutch/Latin
buoy- to float
-ancy condition or state of
The tendency of a body to float or to rise when submerged in a fluid

C

Cadaver
Latin
cadere- to fall or die
-er one that performs that action
A corpse or dead body.

Caddisfly
Old English
cadace- cotton wool (refers to the tube in which the larva lives)
-flēoge fly
Any of various insects with four hairy wings, chewing mouthparts, and long antennae; aquatic larvae.

Caldera
Late Latin
caldaria cooking pot
Large crater formed when the sides of a volcanic cone collapse.

Calendar
Latin
kalendae- account book
-ar relating to or resembling
Any of various systems of reckoning time in which the beginning, length, and divisions of a year are defined.

Calibrate
Arabic
qalib- shoemaker's last
-ate characterized by having
To check, adjust, or determine by comparison with a standard.

Calomel
Greek
kalos- beautiful
-melas black
A tasteless compound, Hg_2Cl_2, used as an insecticide.

Calorie
Latin
calor- heat
Any of several approximately equal units of heat, each measured as the quantity of heat required to raise the temperature of 1 gram of water by 1 degree Celsius from a standard initial temperature.

Calorimeter
Latin/Greek
calor- heat
-meter (metron) instrument or means of measuring; to measure
An apparatus for measuring quantities of absorbed or evolved heat typically generated in a reaction.

Calorimetry
Latin
calor- heat
-metria process of measuring
Measurement of the amount of heat released or absorbed during a chemical reaction.

Calving
Middle English
calve- calf
-ing the act or action of
The process by which a block of a glacier breaks off and falls into the sea to form an iceberg.

Calyx

Greek

kalyx cup

The outer whorl of a flower, the sepals.

Cambium

Latin

cambiare- to exchange

-ium quality or relationship

Plant tissue commonly present as a thin layer that forms new cells on both sides; located either in vascular tissue (vascular cambium), forming xylem on one side and phloem on the other, or in cork (cork cambium or phellogen).

Camouflage

French/Latin

camoufler- to disguise

-age (āticum) **(Latin)** condition or state

Concealment by means of disguise or protective coloring.

Campodeiform

Greek

campo- caterpillar, bend, curve

-dei- god, deity, divine nature

-form having the form of

Applied to insect larvae, grublike, flattened, and elongated with well-developed legs and antennae; many beetle larvae are of this type, as are those of the lacewings.

Canaliculus

Latin

canālis- conduit

-us thing

Very small channels or ducts in the body; normally associated with the Haversian system of compact bone.

Cancer

Latin

cancer crab

A pathological condition marked by the growth and proliferation of neoplastic cells.

Candle

Latin

candela candle

A unit of light intensity equal to the amount of light emitted from a standard source such as a candle or an incandescent light.

Canine

Latin

cani- dog

-ine of or relating to

An animal of the family Canidae; belonging to or characteristic of a dog.

The Heiki Warriors and Natural Selection

Each year on April 24, fishermen who are descendants of the Heike warriors commemorate the last battle of the war between the Heike and Genji samurai clans. On this day, the Heike clan succumbed to its final defeat. The naval battle of Danno-ura was the last stand for this noble clan.

The Heike fought gallantly against an opposing force that greatly outnumbered them. In the end, the survivors, rather than being taken alive, jumped from their ships and committed mass suicide. Among them was their emperor, a seven-year-old boy named Antoku.

The story might have ended there, but for a small group of handmaidens who remained on shore that day. After the war, they lived among the fishermen of the village and bore children.

Over the centuries, the celebration has grown into a legend. The story has it that the Heiki samurai still wander at the bottom of the sea, as evidenced by the many crabs there with markings of what appears to be the face of a samurai.

This is a wonderful example of natural selection. The fishermen of the Danno-ura cast their nets into the inland sea and bring up thousands of crabs. Among them is one with markings vaguely resembling a face on its carapace. The fishermen believe this crab to be sacred and therefore throw it back. The process is repeated countless times. The crabs breed and the likeness of a face is selected for because the crabs bearing it are not harvested. Thus, over time, humans preferentially selected a phenotype, the face of a samurai, to predominate among the population.

Capacitor

Latin

capacitas- spacious

-or person or thing that does something

An electrical circuit element used to store charge temporarily.

Capelin

Latin

cappa- cap or cape

-lin small or little

A small food fish of the smelt family, found in north Atlantic coastal waters.

Capillary
Latin
capill- hairy
-ary pertaining to
As fine or minute as a hair; having a very small bore, as a tube.

Capsid
Latin
cap- catch, seize, take hold of, contain, take, hold
-sid state, condition; having, being, pertaining to, tending to, inclined to
The coating of a protein that encloses the nucleic acid core of a virus.

Capsule
Latin
capsa- box
-ule little, small
A sticky layer that surrounds certain bacteria.

Carapace
Spanish
carapacho covering
The fused chitonous exoskeleton of various invertebrates such as crustaceans.

Carbohydrate
French
carbo- carbon
-hydr- solid compound containing water molecules
-ate characterized by having
Any of a group of organic compounds produced by photosynthetic plants, including sugars, starches, celluloses, and gums, and that serves as a major energy source in the animal diet.

Carbonation
Latin
carbonate- to charge with carbon dioxide gas
-ion state, process, or quality of
Saturation with carbon dioxide gas.

Carcinogen
Greek
karkinos- crab, cancer
-gen to give birth, kind, produce
A substance that induces cancer. Carcinogens are more likely to affect tissues where rapid cellular reproduction takes place.

Carcinoma
Greek
karkinos- crab, cancer
-oma tumor, neoplasm
A malignant growth or tumor.

Impregnating Water with Fixed Air

Joseph Priestley was born in Birstall parish near Leeds, England, in 1733. He was a man of many interests. He was persecuted for his interest in civil rights, government, religion, and philosophy, but it was his sympathetic view of the French people during the French Revolution that led to rumors and conspiracy against him. His home, laboratory, and church in Birmingham were burned to the ground in 1791. He later fled to the United States and took up residence in Northumberland County, Pennsylvania, where he died in 1804.

In 1772 Dr. Joseph Priestley published a paper titled "Impregnating Water with Fixed Air." Here we have the beginnings of carbonated beverages. Priestley experimented with the gas given off by fermenting beer and soon discovered some very interesting characteristics of his collected gas. For example, he learned that the unknown gas was heavier than air, for it remained in the opened containers and did not mix with the ambient air. By performing a common science experiment that is duplicated in most secondary schools across the United States, he came to discover that this gas would extinguish flaming wood chips. The gas that Priestley called "fixed air" was also referred to as "mephitic air" by Joseph Black.

Dr. Priestley's work with "fixed air" led him to perform an experiment where he placed a vessel of water in the gas lingering about the fermented beer. He found that some of the gas dissolved in the standing water, producing a rather tasty beverage, which we know as soda water.

Dr. Priestley's work with gases further led him to the "discovery" of oxygen in 1774. Although oxygen had been identified earlier by Michal Sedziwoj in the sixteenth century and later by Carl Wilhelm Scheele in 1772, Joseph Priestley was the first to publish his results on the gas in 1775, two years before Scheele published his own work. Therefore, Dr. Priestley is commonly credited with the discovery of oxygen.

Cardiac
Greek
kard- heart; pertaining to the heart
-ac pertaining to
Referring to the heart.

Cardialgia
Greek
kard- heart; pertaining to the heart
-algia pain, sense of pain; painful, hurting
Localized pain in the region of the heart.

Cardiology
Greek
kard- heart; pertaining to the heart
-logy (logos) used in the names of sciences or bodies of knowledge
The study of the heart and its actions and diseases.

Cardiomalacia
Greek
kard- heart; pertaining to the heart
-malacia softening of tissue
The softening and degeneration of the walls of the heart, usually because of a disorder.

Cardiomyopathy
Greek
kard- heart; pertaining to the heart
-myo- muscle
-patheia disease; feeling, sensation, perception
A disease or disorder of the heart muscle, especially one of unknown or obscure cause.

Cardiovascular
Greek
kard- heart; pertaining to the heart
-vascul- small vessel
-ar relating to or resembling
Relating to the heart and the blood vessels of the circulatory system.

Carnivore
Latin
caro- meat
-vorare to devour
Any animal that kills and feeds on other animals.

Carotenoid
Latin/Greek
carota- carrot
-oid (oeidēs) resembling, having the appearance of
Any of a class of yellow to red pigments, including the carotenes and the xanthophylls.

Carotid
Greek
karoun- to put to sleep, plunge into sleep or stupor, stupefy
-id state or condition; having, being, pertaining to, tending to, inclined to
Either of the two major arteries, one on each side of the neck, that carry blood to the head; their compression was believed to cause unconsciousness.

Carpal
Greek
carpus- wrist; that which turns
-al of the kind of, pertaining to, having the form or character of
A bone of the wrist; of or relating to the wrist.

Carpel
Greek
karpos fruit
One of the structural units of a pistil, representing a modified, ovule-bearing leaf.

Cartilage
Latin
cartilago- cartilage
-age (āticum) (**Latin**) condition or state
Various tissues containing cartilaginous cells and a matrix composed of water and fibers; it is commonly found in movable joints, the external ear, and the nose, and is the precursor of numerous bones in the human body.

Cartography
Greek
khartes- map, chart, paper
-graphia (graphein) to write, record, draw, describe
The science of map or chart making.

Catabolism
Greek
kata- down, downward; under, lower; against; entirely, completely
-bol- (ballein) to put or throw
-ism state, condition, or quality
Decomposition of larger molecules within cells.

Catadromous
Greek
kata- down, downward; under, lower; against; entirely, completely
-dramein/dromos to run
Refers to fish that migrate from freshwater to the ocean to spawn.

Catalyst
Greek
kata- down, downward; under, lower; against; entirely, completely
-ly- (luein) to loosen, dissolve; dissolution, break
-sis action, process, state, condition
A substance that enables a chemical reaction to proceed, usually at a faster rate or under different conditions than are otherwise possible.

Cataract
Greek
kata- down, downward; under, lower; against; entirely, completely

-arassein to strike
Opacity of the lens or capsule of the eye, causing impairment of vision or blindness.

Catenation
Latin
catena- connection of links or union of parts, as in a chain; a regular or connected series
-ion state, process, or quality of
Bonding of atoms of the same element into chains or rings.

Cathode
Greek
kata- down, downward; under, lower; against; entirely, completely
-hodōs way or road
A negatively charged electrode; an electrolytic cell or a storage battery.

Cation
Greek
kata- down, downward; under, lower; against; entirely, completely
-ion (ienai) to go; something that goes
An ion or group of ions having a positive charge and moving toward the negative electrode in electrolysis.

Caudal
Latin
caud- tail
-al of the kind of, pertaining to, having the form or character of
Constituting, belonging to, or relating to a tail.

Cauterization
Latin
cauter- heat
-ization action, process, or result of doing or making
The process of searing a damaged part of the body by the use of heat or a chemical.

Cecum
Latin
caecus blind
A blind pouch that serves as the entrance to the large intestine.

Ceilometer
Latin
caelum- sky, heaven
-meter (metron) instrument or means of measuring; to measure
A device that measures the height of cloud layers.

Celestial
Latin
caelum- sky, heaven
-ial (variation of *-ia*) relating to or characterized by

Of or relating to the sky or the heavens; planets are celestial bodies.

Cell
Latin
cella chamber
The smallest unit of a living thing that is capable of carrying out all life processes.

Cellulose
Latin
cellula- little cell
-ose sugar
Colorless, insoluble, indigestible polysaccharide that makes up the cell wall.

Celsius
Celsius Swedish scientist (Anders Celsius) who introduced the scale also known as centigrade for measuring temperature
Scale of temperature in which the range from the freezing to the boiling of water is divided into 100 degrees (freezing being 0 and boiling being 100 degrees).

Cenozoic
Greek
kainos- new
-zoe- life
-ic (ikos) relating to or having some characteristic of
Division of geologic time that lasted 65 million years after the Mesozoic.

Centipede
Latin
centi- one hundred
-pede feet
Wormlike arthropods in the class Chilopoda.

Centrifuge
Greek/Latin
kentron- center, sharp point
-fugere to flee
A device for separating components of different densities contained in liquid by spinning at high speed.

Centriole
Greek
kentron- center, sharp point
-ole little
Organelle associated with spindle formation during mitosis in animal cells.

Centripetal (force)
Greek/Latin
kentron- center, sharp point
-petal (petere) moving toward; to seek
The force that opposes the inertia of a body and is required to keep a body in a circular motion.

Centroid
Greek
kentron- center, sharp point
-oid (oeidēs) resembling, having the appearance of
The point in a system of masses each of whose coordinates is a weighted mean of coordinates of the same dimension of points within the system.

Centromere
Greek
kentron- center, sharp point
-mere part of
The area of the chromosome, usually in the center, where sister chromatids are attached.

Centrosome
Greek
kentron- center, sharp point
-soma (somatiko) body
A small region of cytoplasm adjacent to the nucleus that contains the centrioles and serves to organize.

Cephalic
Greek
cephalo- (kephalikos) head
-ic (ikos) relating to or having some characteristic of
Of or relating to the head; anatomical term for "head."

Cephalization
Greek
cephalo- (kephalikos) head
-ization action, process, or result of doing or making
Concentration of sensory and nervous systems in one area of the body, which is called a "head."

Cephalopod
Greek
cephalo- (kephalikos) head
-poda foot
Group of mollusks having a large head, large eyes, prehensile tentacles, and, in most species, an ink sac for protection.

Cephalothorax
Greek/Latin
cephalo- (kephalikos) head
-thorax breastplate, chest
The anterior section of arachnids and many crustaceans, consisting of the fused head and thorax.

Cepheid
Greek
cephalo- (kephalikos) head
-id state, condition; having, being, pertaining to, tending to, inclined to
A variable star that scientists can use to determine how distant a galaxy, or star cluster, is because of its highly regular pulsation.

Ceraceous
Latin
cer- wax
-aceous having the quality of
Waxen, like wax; covered with or resembling wax.

Cercaria
Greek
kerkos- tail
-aria like or connected with
Tadpole-like juveniles of trematodes (flukes).

Cerebellum
Latin
cerebr- of or relating to the brain or cerebrum
-bellum war
A region of brain that lies posterior to the pons and is responsible for voluntary muscular movement, posture, and balance.

Cerebral
Latin
cerebr- of or relating to the brain or cerebrum
-al of the kind of, pertaining to, having the form or character of
The largest part of the brain, consisting of two lobes, the right and left cerebral hemispheres. The cerebrum controls thought and voluntary movement.

Cerebromalacia
Greek
cerebr- of or relating to the brain or cerebrum
-malacia softening of tissue
The abnormal softening of the cerebral parenchyma.

Cerebroside
Latin
cerebr- of or relating to the brain or cerebrum
-ide group of related chemical compounds
A group of lipids that occur most abundantly in the membranes of nerves and brain cells.

Cerussite
Latin
cērussa- a white lead pigment, sometimes used in cosmetics
-ite minerals and fossils
Native lead carbonate; a mineral occurring in colorless, white, or yellowish transparent crystals, with an adamantine, and that is massive and compact.

Cervical
Latin
cervic- stem of cervix
-al of the kind of, pertaining to, having the form or character of
Relating to the neck or any part of the body that resembles a neck.

Cetacean
cetu- whale
-an one that is of, relating to, or belonging to
Order of marine mammals including whales, dolphins, and porpoises.

Chaetotaxy
Greek
chaeto- spine, bristle; long, flowing hair
-taxy arrangement, order; put in order
The arrangement of the bristles or chaetae on an insect, especially important in the classification of the Diptera, Collembla, and several other groups.

Chalcopyrite
Greek
khalkos- copper
-pūr- fire
-ite minerals and fossils
A yellow mineral, essentially $CuFeS_2$, that is an important ore of copper; also called copper pyrite.

Charge
Latin
carrus Gallic type of wagon.
The intrinsic property of matter responsible for all electric phenomena—in particular, for the force of the electromagnetic interaction—occurring in two forms, arbitrarily designated *negative* and *positive*.

Chatoyant
Latin
cattus- cat
-ant performing, promoting, or causing a specified action
A gemstone (cat's-eye) having the capacity of changing its luster or color because of the way narrow bands or streaks of light reflect off its surface.

Cheilostomatoplasty
Greek
cheil- claw, lip, edge, or brim
-stomat- mouth, opening
-plastos- (plassein) something molded; to mold
-y place for an activity, condition, or state
Plastic surgery of the lips and mouth.

Chelicera
Greek
khele- claw
-keras horn
One of a pair of the most anterior head appendages on members of the subphylum Chelicerata.

Cheliped
Greek
khele- claw
-ped foot

A pincerlike claw of a crustacean or arachnid, such as a lobster, crab, or scorpion.

Chemical
Greek
khemeia- chemical; alchemy
-al of the kind of, pertaining to, having the form or character of
A substance composed of chemical elements or one produced by or used in chemical processes.

Chemistry
Greek
khemeia- chemical; alchemy
-metria (metron) the process of measuring
The science of the composition, structure, properties, and reactions of matter, especially of atomic and molecular systems.

Chemoautotroph
Greek
khemeia- chemical; alchemy
-auto- self, same, spontaneous; directed from within
-trophos (trophein) to nourish; food, nutrition; development
Organism that obtains its nourishment through oxidation or inorganic chemical compounds.

Chemoheterotroph
Greek
khemeia- chemical; alchemy
-heteros- different
-trophos (trophein) to nourish; food, nutrition; development
Any of a group of bacteria that, in addition to deriving energy from chemical reactions, synthesize all necessary organic compounds from carbon dioxide.

Chemotherapy
Latin
khemeia- chemical; alchemy
-therapeuein heal, cure; treatment
A treatment for cancers that involves administering chemicals that are toxic to malignant cells.

Chiasma
Greek
khiazein to mark with an X
In anatomy, the crossing or intersecting of two tracts; the optic chiasma. In genetics, the point of contact between paired chromatids.

Chilopoda
Greek
kheilos- lip
-poda foot
A very large group of insects that includes centipedes; they are characterized by having elongated legs attached to each segment, with a pair of legs

in the thorax that serve as fangs, and by having very powerful mouthparts.

Chimera
Greek
chimaira she-goat
An organism composed of two or more genetically distinct tissues, such as one that is partly male and partly female, or an artificially produced individual having tissues of several species.

Chiropractic
Greek
chir- hand; pertaining to the hand or hands
-praktikos- practical
-ic (ikos) relating to or having some characteristic of
A system of therapy in which disease is considered the result of abnormal function of the nervous system; treatment usually involves manipulation of the spinal column and other body structures.

Chiroptera
Greek
chir- hand; pertaining to the hand or hands
-pteron wing
Order of flying mammals (bats).

Chloragogen
Greek
chlor- the color green, yellow-green, or light green
-agogos- a leading, a guide
-gen to give birth, kind, produce
Modified greenish or brownish peritoneal cells clustered around the digestive tract of certain annelids; they apparently aid in the elimination of nitrogenous wastes and in food transport.

Chlorofluorocarbon
Greek
chlor- the color green, yellow-green, or light green
-fluere- chemical element; to flow
-carbo- coal, charcoal
-on a particle
Any of several simple gaseous compounds that contain carbon, chlorine, fluorine, and sometimes hydrogen.

Chloroform
Greek/Latin
chlor- the color green, yellow-green, or light green
-formyl [*-form(ic)* found in ants + *yle* wood, matter]
A clear, colorless, sweet-smelling liquid used in refrigerants, propellants, and resins; as a solvent; and sometimes as an anesthetic.

Chlorophyll
Greek
chlor- the color green, yellow-green, or light green
-phullon leaf
Green pigment found in photosynthetic organisms that is capable of absorbing light and converting it to energy from oxidation and reduction in the photosynthesis of carbohydrates.

Chloroplast
Greek
chlor- the color green, yellow-green, or light green
-plastos (plassein) something molded; to mold
Chlorophyll-containing plasmid found in algal and green plants.

Choanoblast
Greek
choane- funnel
-blastos bud, germ cell
A cell that gives rise to one or more collar bodies, especially in the sponge class Hexactinellida.

Choanocytes
Greek
choane- funnel
-cyte (kutos) sac or bladder that contains fluid
One of the flagellate collar cells that line the cavities and canals of sponges.

Cholecystectomy
New Latin
khole- bile, gall
-kustis- (cyst) sac or bladder that contains fluid
-ekt- outside, external, beyond
-tomos (temnein) to cut, incise, section
Surgical excision of the gallbladder.

Cholelith
Greek
khole- bile, gall
-lith stone, rock
A small, hard pathological concretion composed chiefly of cholesterol, calcium salts, and bile pigments, formed in the gallbladder or in a bile duct; gallstone.

Cholesterol
Greek
khole- bile, gall
-steros- solid
-ol chemical derivative
A white crystalline substance found in animal tissues and various foods that is normally synthesized by the liver and is important as a constituent of cell membranes and a precursor to steroid hormones.

Chondroblast
Greek
khondros- granule, cartilage
-blastos bud, germ cell
An immature cartilage cell found in growing cartilage.

Chondroclast
Greek

khondros- granule, cartilage
-klastos break, break in pieces
A cartilaginous cell involved with the reabsorbtion of the cartilaginous matrix.

Chondrocyte
Greek
khondros- granule, cartilage
-cyte (kutos) sac or bladder that contains fluid
A mature cartilage cell that can be found in the lacunae of the cartilaginous matrix.

Chondromalacia
Greek
khondros- granule, cartilage
-malacia softening of tissue
Softening of any cartilage, usually because of a physiological disorder.

Chordate
Greek
khorde- gut, string of a musical instrument
-ate characterized by having
Of, pertaining to, or belonging to the phylum Chordata or to a chordate subphylum; animals having at least at some stage of development of a notochord, a dorsally situated central nervous system, and gill clefts.

Choroid
Greek
khorion- afterbirth
-oid (oeidēs) resembling, having the appearance of
The very dark brown vascular coat found between the sclerotic coat and the retina of the eye.

Chromatics
Greek
khromat- color
-ic (ikos) relating to or having some characteristic of
The scientific study of color.

Chromatid
Greek
khromat- color
-id state, condition; having, being, pertaining to, tending to, inclined to
One of the two identical threadlike filaments of a chromosome.

Chromatin
Greek
khromat- color
-in protein or derived from protein
A complex of nucleic acids and proteins in the cell nucleus that stains readily with basic dyes and condenses to form chromosomes during cell division.

Chromatography
Greek

khromat- color
-graphia (graphein) to write, record, draw, describe
Analysis of mixtures of chemical compounds by passing solutions of them through an absorbent.

Chromogen
Greek
khromat- color
-gen to give birth, kind, produce
A substance capable of conversion to a pigment or dye.

Chromophore
Greek
khromat- color
-phore bearer, carrier
A chemical group capable of selective light absorption resulting in the coloration of certain organic compounds.

Chromosome
Greek
khromat- color
-soma (somatiko) body
Any one of the threadlike nucleoprotein structures in the nucleus of the cell that function in the transmission of genetic information.

Chromosphere
Greek
khromat- color
-sphaira a globe shape, ball, sphere
An incandescent, transparent layer of gas lying above and surrounding the photosphere of the sun.

Chronic
Greek
khronos- time
-ic (ikos) relating to or having some characteristic of
Lasting a long time, long-continuing, lingering, inveterate (as diseases).

Chronobiology
Greek
khronos- time
-bios- life, living organisms or tissue
-logy (logos) used in the names of sciences or bodies of knowledge
The scientific study of the effect of time on living systems.

Chronogram
Greek
khronos- time
-gram something written or drawn; a record
The record produced by a chronograph.

Chronometry
Greek

Eratosthenes' Shadows

"Let none enter here who are ignorant of geometry." This quote was inscribed above the entrance of Plato's school, illustrating the importance of mathematics to the early philosopher-scientists of Greece and Egypt. Without knowledge of geometry, we'd be left with many elegant theories, perhaps, but no reasoned explanations. Plato, though not a mathematician, understood this.

This brings us to Eratosthenes (276–194 BC), born in what is now Libya. A man of considerable influence, Eratosthenes was a mathematician, astronomer, geographer, poet, historian, and philosopher. He studied and worked, probably as a director, in the Great Library of Alexandria. It is here he read that at noon every June 21, the sun cast no shadow in the Egyptian village of Syene. And on that same day at the same hour, the full face of the sun was reflected in the waters of the village's deep well. To even the uninformed observer, it was obvious that the sun was directly overhead.

Perhaps out of curiosity or an attempt to validate the account of Syene, Eratosthenes, using only a stick placed in the sand at Alexandria (a considerable distance north of Syene), made the observation that at noon of June 21, a rather lengthy shadow was cast. Undoubtedly, Eratosthenes asked himself what possibly could account for such a phenomenon. If the earth were flat, like the maps, then the shadows should be the same length—provided, of course, that the sun was a considerable distance from the earth. Or could the earth be a sphere, and not flat at all? Knowing that the distance from Alexandria to Syene was about 800 kilometers, and observing and calculating the difference between the shadow lengths at the two locations, Eratosthenes calculated that the degree of the angle where the sticks would intersect deep within the earth was probably close to 7 degrees. Having that bit of information, he was able to determine the circumference of the earth. If the opposite side of a 7-degree angle is 800 kilometers, and there are 360 degrees in a circle, the resulting circumference is around 40,000 kilometers. He was pretty accurate for someone using only his intellect and no technology.

khronos- time
-metria (metron) the process of measuring
The scientific measurement of time.

Chrysalis
Greek
khrūsallid gold-colored pupa of a butterfly
The protective stage of development in moths and butterflies in which the pupa is contained in a tough case or cocoon.

Chyle
Greek
chylos juice
A milky fluid containing emulsified fat and other products of digestion formed from the chyme in the small intestine and conveyed by the lacteals and the thoracic duct to the veins.

Cilia
Latin
cili- a small hair
-ia names of diseases, place names, or Latinizing plurals
Small hairlike projections that help ciliates move, sense their environment, and collect food.

Ciliate
Latin
cili- a small hair
-ate characterized by having

Any of a group of animal-like protists that are characterized by having cilia.

Circadian
Latin
circum- around
-diurnus- day
-an one that is of, relating to, or belonging to
Designating physiological activity that occurs approximately every twenty-four hours, or the rhythm of such activity.

Circuit
Greek
kirkos circle
A set of electronic components that perform a particular function in an electronic system.

Circular
Latin
circulus- to make circular
-ar relating to or resembling
Referring to a path that follows the shape of a circle.

Circulation
Latin
circulus- to make circular
-ion state, process, or quality of
Movement or flow through a circle or circuit.

Circumcision
Latin
circum- in a circle; around, about, surrounding
-caedere- to cut
-ion state, process, or quality of
The act of cutting around; the cutting and removal of all of the prepuce in males or the prepuce, clitoris, or labia in females.

Circumference
Greek
circum- in a circle; around, about, surrounding
-ferre to carry
The boundary line of a circle, or the length of such a boundary.

Circumlunar
Latin
circum- in a circle; around, about, surrounding
-lunar moon, light, shine
Revolving around or surrounding the moon.

Cirque
French (from Latin)
circus circle
A steep, hollow, bowl-shaped basin occurring at the upper end of a mountain valley.

Cirrhosis
Greek
kirrhos- tawny yellow
-sis action, process, state, condition
A chronic disease of the liver characterized by the replacement of normal tissue with fibrous tissue, the loss of functional liver cells, and an abnormal yellowish appearance.

Cirrus
Latin
cirro hair; wispy
High clouds with a base of 6,000 meters.

Cistron
Latin
cist- to cut
-on a particle
Segment of DNA that is required in order to synthesize a complete polypeptide chain.

Cladistics
Greek
klados- branch or sprout
-ic (ikos) relating to or having some characteristic of
A system of arranging taxa to reflect phylogenetic relationships.

Cladogram
Greek
klados- branch or sprout
-gramma letter

A branching diagram showing the pattern of sharing evolutionarily derived characters among species or higher taxa.

Clastic
Greek
klastos- broken
-ic (ikos) relating to or having some characteristic of
Sedimentary rock formed by fragments of previously existing rock.

Clavicle
Latin
clāvis- key (from its shape)
-ic (ikos) relating to or having some characteristic of
One of two slender, key-shaped bones located be-tween the scapula and the manubrium of the sternum.

Cleavage
Middle English
cleave- to split or separate
-age (āticum) (**Latin**) condition or state
Splitting or separation along a natural Zline of division.

Clepsydra
Greek
kleps- to steal
-hudor water
An ancient device used for measuring time by the dripping of water from a graduated vessel.

Climate
Greek
klime- slope
-ate characterized by having
General conditions of temperature and precipitation for an area over a period of time.

Clinarthrosis
Greek
klinein- to lean; sloping
-arthr- pertaining to the joints
-osis process, condition, or state of
Abnormal deviation in the alignment of the bones at a joint.

Cline
Greek
klinein to lean; sloping
A continuous series of differences in structure or function exhibited by the members of a species along a line extending from one part of their range to another.

Clinic
Greek
klinikos- pertaining to a bed or couch
-ic (ikos) relating to or having some characteristic of

A clinical lecture; examination of patients before a class of students; instruction at the bedside.

Clinician
Greek

klinikos- pertaining to a bed or couch
-an one that is of, relating to, or belonging to
An experienced practitioner such as a nurse, physician, or psychologist as opposed to someone involved in research.

Clinicopathologic
Greek

klinikos- pertaining to a bed or couch
-pathos- feeling, sensation, perception; suffering, disease
-logic talk, speak; speech; word
Pertaining both to the symptoms of a disease and to its pathology.

Clinocephaly
Greek

klinikos- pertaining to a bed or couch
-cephaly (kephalikos) head
Congenital flatness or concavity of the vertex of the head.

Clinodactyly
Greek

klinein- to lean; sloping
-dactylos finger, toe
Permanent lateral or medial deviation or deflection of one or more fingers.

Clinography
Greek

klinikos- pertaining to a bed or couch
-graphia (graphein) to write, record, draw, describe
A system of graphical representations of the temperature, symptoms, and pathological manifestations exhibited by a patient.

Clinoid
Greek

klinikos- pertaining to a bed or couch
-oid (oeidēs) resembling, having the appearance o
Bed-shaped, as the clinoid processes of the sphenoid bone.

Clinostatism
Greek

klinikos- pertaining to a bed or couch
-statos- standing, stay, make firm, fixed, balanced
-ism state, condition, or quality
The condition of lying down or being in the horizontal position.

Cliseometer
Greek

How Do You Discover the Invisible?

It has been said that Empedocles of Agrigentum (ca. 490–430 BC), a mystic, poet, and physician, was so self-absorbed that he considered himself a god and was perhaps considered divine by others. Empedocles postulated that all matter is made up of four "roots": water, earth, fire, and air. He declared that love (*phila*) was the force that held these roots together and that discord (*neikos*) was the force at work to keep them apart.

We know air to be an invisible medium, but to the ancient Greeks, the wind was the breath of the gods. It had no substance and no tangible qualities. How, then, could Empedocles prove the existence of air? One of the rare Greek scientists who actually did experiments, Empedocles used a clepsydra, a common household ladle or "water clock," for his test. A clepsydra was a vessel with markings and one or more small holes at its base to allow water to drip out. The top of the vessel had a strawlike tube attached. When Empedocles filled the clepsydra with water, it dripped out the bottom. But when he put his finger over the opening of the tube at the top of the vessel, the water stopped dripping. When he tried filling the vessel with his thumb over the opening of the tube, as he submerged the clepsydra, no water could enter the vessel through the other end. What could be causing this? Empedocles argued that something invisible but with substance (matter) filled the void in the vessel. If it could not be moved out, then nothing could take its place. Hence air, though invisible, exists and has substance.

---

klisis- inclination
-meter (metron) instrument or means of measuring; to measure
An instrument for measuring the angle that the pelvic axis makes with the spinal column.

Clitellum
Latin

clitellae- packsaddle
-um (**singular**) structure
-a (**plural**) structure
A thickened glandular section of the body wall of some annelids that secretes a viscid sac in which the eggs are deposited.

Clitoris
Greek
kleitoris clitoris
An organ of very sensitive tissue located just anterior to the urinary meatus.

Cloaca
Latin
cloa'cae drain
A common passage for fecal, urinary, and reproductive discharge in monotremes, birds, and lower vertebrates.

Clone
Greek
klōn young shoot or twig
A cell, group of cells, or organism that is descended from and genetically identical to a single common ancestor, such as a bacterial colony whose members arose from a single original cell.

Clonogenic
Greek
klōn- young shoot or twig
-gen- to give birth, kind, produce
-ic (ikos) relating to or having some characteristic of
An organism arising from or consisting of a clone of cells.

Clupeine
Latin
clupea- herring, small fish
-ine in a chemical substance
A protamine obtainable from the spermatozoa of the herring.

Cnemitis
Greek
knēmē- leg
-itis inflammation, burning sensation
Inflammation of the tibia.

Cnemoscoliosis
Greek
knēmē- leg
-scoli- curvature; curved, twisted, crooked
-sis action, process, state, condition
A lateral bending of the lower limb.

Cnicus
Greek
knēkos- safflower
-us thing
A genus of European herbs of the family Compositae.

Cnidaria
Greek
kin' dh- to sting; nettle
-ia names of diseases, place names, or Latinizing plurals
Phylum consisting of organisms with special stinging cells.

Cnidoblast
Greek
kin' dh- to sting; nettle
-blastos bud, germ cell
The epidermal cells of coelenterates that contain the nematocysts, especially numerous on the tentacles.

Cnidocil
Greek
kin' dh- to sting; nettle
-cilium hair
Triggerlike spine on a nematocyst.

Cnidocilium
Greek
kin' dh- to sting; nettle
-cili- a small hair
-um (**singular**) structure
-a (**plural**) structure
A bristle-like process at one end of a cnidoblast, which, when stimulated, triggers the discharge of the nematocyst.

Cnidocytes
Latin
kin' dh- to sting; nettle
-cyte (kutos) sac or bladder that contains fluid
Stinging cell used by cnidarians to stun their prey.

Coacervate
Greek
co- together, with
-acervāre- to heap
-ate of or having to do with
The viscous phase separating from a colloid-containing system in the phenomenon of coacervation.

Coacervation
Greek
co- together, with
-acervāre- to heap
-ion state, process, or quality of
The separation of a mixture of two liquids, one or both of which are colloids, into two phases; one (the coacervate) contains the colloidal particles, and the other is an aqueous solution (e.g., as when gum arabic is added to gelatin).

Coadunation
Latin
co- together, with
-unus- one
-ion state, process, or quality of
Union of dissimilar substances in one mass.

Coagulate

Latin
co- together, with
-agulum- to condense; to drive
-ate of or having to do with
To cause the transformation of a liquid into a soft, semisolid, or solid mass.

Coalescence

Latin
co- together, with
-alescere- to come together or grow
The act of growing together; the act of uniting.

Coccidium

Greek
co- together, with
-kokkos- berry, grain, seed
-ium quality or relationship
In former systems of classification, a genus of coccidians, the organisms of which have been assigned to other genera.

Cochlea

Greek
kokhlias snail
A spiral-shaped cavity of the inner ear that contains nerve endings essential for hearing.

Codominance

Latin
co- together, with
-domo- house, home
-ance state, quality
In genetics, the tendency of certain (dominant) alleles to mask the expression of their corresponding (recessive) alleles.

Codominant

Latin
co- together
-dominae to rule
Referring to an equal degree of dominance of two alleles or traits fully expressed in a phenotype.

Codon

Latin
cod- a code of laws; a writing tablet; an account book
-on subatomic particle
A group of three nucleotides that specifies the addition of one of the twenty amino acids during translation of an mRNA into a polypeptide. Strings of codons form genes, and strings of genes form chromosomes.

Coefficient

Latin
co- together, with
-efficiens- efficient
-ent causing an action; being in a specific state; within
Number that serves as a measure of some property or characteristic; numerical factor by which the value of another is multiplied.

Coelenterata

Greek
koilos- hollow cavity
-enteron intestine
Former name for a phylum of marine invertebrates including sea anemones, hydras, jellyfish, and corals, which are now assigned to the phylum Cnidaria.

Coelenteron

Greek
koilos- hollow cavity
-enteron intestine
Internal cavity of a cnidarian; gastrovascular cavity; archenteron.

Coelom

Greek
koilos hollow cavity
The epithelium-lined space between the body wall and the digestive tract of metazoans above lower worms.

Coelomoduct

Greek
koilos- hollow cavity
-ductus leading
A duct that carries gametes or excretory products (or both) from the coelom to the exterior.

Coenocytic

Greek
coeno- shared
-kutos- (cyto) sac or bladder that contains fluid
-ic (ikos) relating to or having some characteristic of
Multinucleate, with nuclei not separated by cross walls.

Cohesion

Latin
co- together, with
-haerere- to stick together
-ion state, process, or quality of
The binding together of like molecules.

Cohesive

Latin
co- together, with
-haerere- to stick together
-ive performing an action
Holding the particles of a homogeneous body together.

Coitus
Latin
co- together, with
-ire to go, come
The sexual union of a male and female.

Colchicine
Latin
kolkhikon- meadow saffron
-ine of or relating to
Poisonous, pale-yellow alkaloid that inhibits mitosis.

Cold
Middle English
caeld cold
In weather, having a low atmospheric temperature. In life science, a common name for infections of the upper respiratory system.

Coleoptera
Greek
koleos- sheath
-pteron wing
Insect order having an anterior pair of hard and horny wings covering a softer pair of posterior wings, and two pairs of jaws adapted for feeding; beetles, weevils.

Coleoptile
Greek
koleos- sheath
-ptilon plume
A protective sheath enclosing the shoot tip and embryonic leaves of grasses.

Collagen
Greek
kolla- glue
-gen to give birth, kind, produce
A tough, fibrous protein occurring in vertebrates as the chief constituent of collagenous tissue, and also occurring in invertebrates—for example, in the cuticle of nematodes.

Collembola
Greek
kolla- glue
-mbolon wedge, peg
Springtail; minute insect that lacks wings and has a ventral tube, or collophore, on the first abdominal segment and an abdominal forked furcula, or spring used to propel the organism forward.

Collenchyma
Greek
col- with, together
-khumos juice
Tissues that provide mechanical support in many young, growing plant structures (stems, petioles, leaves) but are uncommon in roots.

Collencyte
Greek
kolla- glue
-cyte (kutos) sac or bladder that contains fluid
A type of cell in sponges that secretes fibrillar collagen.

Colligative
Latin
com- together, with; joint; jointly
-ligare- to tie, bind
-ive performing an action
Depending on the quantity of molecules but not on their chemical nature.

Colloblast
Greek
kolla- glue
-blastos bud, germ cell
A glue-secreting cell on the tentacles of ctenophores.

Colloid
Greek
kolla- glue
-oid (oeidēs) resembling, having the appearance of
A suspension of final divided particles in a continuous medium.

Collophore
Greek
kolla- glue
-phore bearer, carrier
A suckerlike organ at the base of the abdomen of insects belonging to Collembola (springtails).

Colon
Greek
kolon large intestine
The section of the large intestine extending from the cecum to the rectum.

Combustion
Latin
com- (con) together, with, jointly
-bustus- to burn
-ion state, process, or quality of
A chemical process accompanied by the evolution of light and heat.

Comet
Greek
kometes long-haired
A celestial body in an elliptical orbit around the sun; a brightly illuminated mass composed of ice and rock and displaying a long, glowing tail when its orbit takes it near the sun.

Commensalism
Latin
com- (con) together, with, jointly
-mensa- table
-ism state, condition, or quality
A relationship between organisms where one benefits while the other is unaffected; sharing a meal.

Commissure
Latin
com- (con) together, with, jointly
-mittere to put
A point or line of union or junction, especially between two anatomical parts, such as the tract of nerve fibers passing from one side to the other of the spine or brain.

Community
Latin
communis- commons
-ity state or quality of
All of the populations of all species existing together within an ecological system.

Competition
Latin
com- (con) together, with, jointly
-peter- to strive
-ion state, process, or quality of
The struggle for existence among organisms.

Complex
Latin
com- (con) together, with, jointly
-plexus an embrace
A group of items, such as chemical molecules, that are related in structure or function.

Component
Latin
com- (con) together, with, jointly
-ponere- to put together
-ent causing an action; being in a specific state; within
Unit resulting from the subdivision of a vector into axial parts.

Compound
Latin
com- (con) together, with, jointly
-ponere to put
A pure substance that is composed of two or more elements in fixed proportions and that can be chemically decomposed into these elements.

Compression
Latin
com- (con) together, with, jointly
-premere- to press
-ion state, process, or quality of
An increase in the density of something as a result of compacting.

Concave
Latin
com- (con) together, with, jointly
-cavare to make hollow
Curved like the interior of an arched circle.

Concentric
Latin
com- (con) together, with, jointly
-centrum center
Describing circles within circles, with the system having a common center.

Conchoidal
Greek
conch- shell
-id- state, condition; having, being, pertaining to
-al of the kind of, pertaining to, having the form or character of
Of, relating to, or being a surface characterized by smooth, shell-like convexities and concavities, as on fractured obsidian.

Concurrent
Latin
com- (con) together, with, jointly
-currere to coincide
Happening at the same time or operating in conjunction with one another.

Condensation
Latin
com- (con) together, with, jointly
-dens- to press close together
-ion state, process, or quality of
The process by which a gas changes to a liquid.

Conduction
Latin
com- (con) together, with, jointly
-ducere- to bring together
-ion state, process, or quality of
The flow of electron through a material to produce electric current.

Conductive
Latin
com- (con) together, with, jointly
-ducere- to bring together
-ive performing an action
Exhibiting the power or ability to conduct or transmit heat, electricity, or sound.

Conductor
Latin
com- (con) together, with, jointly
-ducere- to bring together
-or person or thing that does something
A substance or medium that conducts heat, light, sound, or especially an electrical charge.

Congenital
Latin
com- (con) together, with, jointly
-genitus- born; to bear
-al of the kind of, pertaining to, having the form or character of
Of or relating to a condition that is present at birth.

Conidiophore
Greek
konis- dust
-phore bearer, carrier
A specialized fungal form that asexually produces conidial spores.

Conidium
Greek
konis dust
An asexually produced fungal spore, formed on a conidiophore.

Conifer
Greek
konos- cone
-ferre to bear
Any of an order of mostly evergreen trees and shrubs with true cones and others (such as yews) with an arillate fruit.

Coniferous
Latin
konos- cone
-ferre- to bear
-ous full of, having the quality of, relating to
Relating to the groups of plants that bear cones (pines and cypress).

Coniine
Greek
koneion- poison hemlock
-ine a chemical substance; of or relating to
A poisonous, colorless liquid alkaloid found in poison hemlock.

Conjugation
Latin
com- (con) together, with, jointly; compress, converge
-jugare- to join together
-ion state, process, or quality of
The joining of unicellular organisms to exchange hereditary material.

Conjunctiva
Latin
com- (con) together, with, jointly; compress, converge
-jungere- to join
-iva of the quality of; tending to, inclined to
The mucous membrane that lines the inner surface of the eyelid and the exposed surface of the eyeball.

Conodont
Greek
konos- cone
-odontos tooth
Toothlike element from a Paleozoic animal now believed to have been an early marine vertebrate.

Conscious
com- (con) together, with, jointly; compress, converge
-scire- to know
-ous full of, having the quality of, relating to
Being aware and having perception of one's own existence, sensations, and thoughts and of the surrounding environment.

Conservation
Latin
com- (con) together, with, jointly; compress, converge
-servare- to preserve
-ion state, process, or quality of
The process of protecting, preserving, and using wisely the natural resources.

Constant
Latin
com- (con) together, with, jointly; compress, converge
-stare to stand firm
A numerical value that does not change.

Constellation
Latin
com- (con) together, with, jointly; compress, converge
-stella- star
-ion state, process, or quality of
A group of stars that form a pattern.

Constipation
Latin
com- (con) together, with, jointly; compress, converge
-stipare- to press together
-ion state, process, or quality of
Infrequent and difficult movement of bowels.

Constrictor
Latin
com- (con) together, with, jointly; compress, converge
-stingere- to pull
-or condition or property of things or persons; person who does something
A muscle that contracts a cavity or orifice or compresses an organ.

Consumer
Latin
com- (con) together, with, jointly; compress, converge
-sumere- to take
-er one that performs an action
Any organism that is incapable of producing its own food by photosynthesis or chemosynthesis; it derives its nutrients through the consumption of producers or other consumers.

Contagious
Latin
com- (con) together, with, jointly; compress, converge
-teg- touch, reach, handle
-ous full of, having the quality of, relating to
Transmissible by direct or indirect contact; capable of transmitting disease; spreading or tending to spread from one to another; infectious.

Continent
Latin
com- (con) together, with, jointly; compress, converge
-tenere- to hold together
-ent causing an action; being in a specific state; within
One of the principal land masses of the earth.

Contour
Latin
com- (con) together, with, jointly; compress, converge
-tornāre to round off
Feathers that make up general outline of a bird.

Contusion
Latin
com- (con) together, with, jointly; compress, converge
-tundere- to beat
-ion state, process, or quality of
An injury in which the skin is not broken, often characterized by ruptured blood vessels and discoloration; a bruise.

Convection
Latin
com- (con) together, with, jointly; compress, converge
-vehere- to carry
-ion state, process, or quality of
Transfer of energy by the flow of a heated substance.

Conversion
Latin
com- (con) together, with, jointly; compress, converge
-vertere- to turn around
-ion state, process, or quality of
The process in which something is changed from one use, function, or purpose to another.

Convex
Latin
com- (con) together, with, jointly; compress, converge
-vextus to be vaulted
Having a surface that curves outward.

Copepod
Greek
kope- oar
-pod foot
Any of numerous minute marine and freshwater crustaceans of the subclass Copepoda, having an elongated body and a forked tail.

Coprophagy
Greek
kopros- dung
-phagei- to eat
-y place for an activity; condition, state
Feeding on dung or excrement as a normal behavior among animals; reingestion of feces.

Cornea
Latin
corneus horny
The outer transparent, convex part of the front of the eyeball; it covers the iris and the pupil of the eye.

Corniculate
Latin
corniculum horn, hornlike structure
-ate of or having to do with
Bearing or furnished with one or more small horns.

Corolla
Latin
corolla small garland
Whorl of a flower that consists of the petals.

Corona
Latin
corona crown
The luminous, irregular envelope of highly ionized gas outside the chromosphere of the sun.

Coronary

Latin

corona- crown

-ary of, relating to, or connected with

Of, relating to, or being the coronary arteries or coronary veins; of or relating to the heart.

Corrugator (supercilii)

Latin

com- (con) together, with, jointly; compress, converge

-rigare- to wrinkle

-or a condition or property of things or persons

A muscle of the eyelid, located under the eyebrow, functioning to draw the eyebrow downward and inward, wrinkling the adjacent skin.

Cortex

Latin

cortic bark, rind, that which is stripped off

The outer layer of an internal organ or body structure, as of the kidney or adrenal gland; the outer layer of gray matter that covers the surface of the cerebral hemisphere.

Cosmic

Greek

kosmos universe

Of or relating to the universe, especially as distinct from earth.

Cosmochemistry

Greek

kosmos- universe, order

-khemeia- chemical; alchemy

-y place for an activity, condition, or state

The science of the chemical composition of the universe.

Cosmogony

Greek

kosmos- universe, order

-gonos offspring

The astrophysical study of the origin and evolution of the universe.

Cosmology

Greek

kosmos- universe, order

-logy (logos) used in the names of sciences or bodies of knowledge

The study of the physical universe considered as a totality of phenomena in time and space.

Costalgia

Latin

costo- rib

-algia pain, sense of pain; painful, hurting

Plueritic pain in the chest.

Costocervical

Latin

costo- rib

-cervic- stem of cervix

-al of the kind of, pertaining to, having the form or character of

Concerning the ribs and the neck.

Costoinferior

Latin

costo- rib

-inferus below, low

Relating to the lower rib.

Costophrenic

Latin

costo- rib

-phren- diaphragm, midriff, heart

-ic (ikos) relating to or having some characteristic of

Referring to the ribs and diaphragm.

Costopneumopexy

Latin

costo- rib

-pneumon- wind, breath

-pexy attaching; surgical fixation of an organ

The surgical anchoring of a lung to a rib.

Costosuperior

Latin

costo- rib

-superus higher, upper

Relating to the upper rib.

Costotome

Latin

costo- rib

-tomos (temnein) to cut, incise, section

An instrument designed to cut through ribs.

Cotyledon

Greek

kotuledon a kind of plant; a seed leaf; a hollow or cup-shaped object

The one or two seed leaves of an angiosperm embryo.

Coumarin

Portuguese

cumaru- tonka bean tree

-in neutral chemical; protein derivative

A fragrant crystalline compound extracted from several plants and widely used in perfumes.

Couple

Latin

copula bond or pair

A pair of forces of equal magnitude acting in parallel but opposite directions.

Covalence
Latin
co- to the same extent or degree; together, jointly
-valere to be strong
The number of electron pairs an atom can share with other atoms.

Covariant
Latin
co- to the same extent or degree; together, jointly
-variare to vary
Expressing or relating to the principle that physical laws have the same form regardless of the coordinate system in which they are expressed.

Coxopodite
Latin
coxa- hip
-podos- foot
-ite component of a part of the body
The proximal joint of an insect or arachnid leg; in crustaceans, the proximal joint of the protopod.

Cracking
Middle English
cracian- to break apart
-ing the act of
Thermal decomposition of a complex substance.

Craniomalacia
Greek
kranion- skull
-malacia softening of tissue
Softening of the bones of the skull.

Cranium
Greek
kranion skull
The part of the skull that encloses the brain.

Crater
Greek
krater bowl for mixing wine and water
Funnel-shaped pit or depression at the top of a volcanic cone.

Creatinine
Greek
kreat- flesh
-ine a chemical substance
A waste product of protein usage in cells; nitrogenous wastes excreted in urine.

Cremaster
Latin
crem- to hang; hung, hung up
-ster one that is associated with, participates in, makes, or does
The hooklike process on the end of a chrysalis that attaches the pupa to the stem or twig, for example.

Crepuscular
Latin
creper- dark
-ar relating to or resembling
In biology, relating to organisms that become active after twilight (e.g., bats).

Cretaceous
Latin
creta- chalk
-eous full of, having the quality or nature of, relating to
The final period of the Mesozoic era, spanning the time between 145 and 65 million years ago.

Crevasse
French
crevace crevice
A deep fissure; a chasm.

Crocodile
Greek
kroke- pebble
-drilos circumcised man; worm
The name given to various large aquatic reptiles found in the tropics and subtropics with thick, bumpy skin and long, tapered jaws.

Crop
Old English
cropp craw
A pouched enlargement of the gullet that serves as a receptacle for food and for its preliminary maceration.

Crust
Latin
crusta shell, hard surface of a body
The outermost layer of the earth's surface, extending downward about 20 miles on the land masses and 3 to 10 miles down beneath the ocean floor.

Crustacean
Latin
crusta- shell, hard surface of a body
-acean belonging to a taxonomical group
One of the classes of the phylum Arthropoda possessing shells.

Cryptobiotic
Greek
kryptos- hidden
-bios- life, living organisms or tissue
-ic (ikos) relating to or having some characteristic of
Living in concealment; refers to insects and other animals that live in secluded situations, such as underground or in wood, and also to tardigrades and some nematodes, rotifers, and others that survive harsh environmental conditions by assuming for a time a state of very low metabolism.

Crystal
Latin
krustallos- ice, crystal; freeze; icelike
-al of the kind of, pertaining to, having the form or character of
Very clear glass; a homogeneous solid formed by a repeating three-dimensional pattern.

Crystalline
Greek
krustallos- ice, crystal; freeze; icelike
-ine of or relating to
Resembling crystal, as in transparency or distinctness of structure or outline.

Crystallization
Greek
krustallos- ice, crystal; freeze; icelike
-ion state, process, or quality of
The process of forming solid crystals in solution due to the solute solubility exceeding that of the solvent.

Culture
Latin
cult- to care for; to dwell, to inhabit
-ura act, process, condition
The growing of microorganisms, tissue cells, or other living matter in a specially prepared nutrient medium.

Cumulonimbus
Latin
cumul- pile or heap
-nimbus cloud
An extremely dense, vertically developed cumulus with a glaciated top extending to great heights.

Cumulus
Latin
cumul- pile or heap
-us thing
Heap, Pile, or mass.

Cuspid
Latin
cuspis- sharp point, cusp
-id state or condition; having, being, pertaining to, tending to, inclined to
Pointed or conical teeth, usually referring to the canine teeth.

Cuticle
Latin
cutis skin
A waxy layer that coats the surface of stems, leaves, and other plant parts exposed to air.

Cutoff
Old English
cutten- to separate into parts with or as if with a sharp-edged instrument
-of no longer taking place; canceled
A new channel cut by a river across the neck of an oxbow.

Cyanobacteria
Greek
cyano- (kyanos) blue, dark blue
-baktron- staff, rod
-ia names of diseases, place names, or Latinizing plurals
Microscopic, photosynthetic prokaryotes that formed stromatilites and changed the earth's atmosphere by producing oxygen.

Cyanoderma
Greek
cyano- (kyanos) blue, dark blue
-derma skin
Bluish discoloration of the skin.

Cyanosis
New Latin
cyano- (kyanos) blue, dark blue
-sis action, process, state, condition
Bluish discoloration of the skin due to deficient oxygenation of the blood.

Cycads
Greek
cyc- (koïx) a kind of palm tree, perhaps of Egyptian origin
-ad member of a botanical group
Any of an order (Cycadales) of dioecious cycadophytes that are represented by a single surviving family (Cycadaceae) of palmlike tropical plants that reproduce by means of spermatozoids.

Cyclase
Greek
kyklos- circle, wheel, cycle; rotate
-ase indicating an enzyme
Enzyme that forms a cyclic compound.

Cycle
Greek
kyklos circle, wheel, cycle, rotate
An interval of time during which a sequence of a recurring events or phenomena is completed.

Cycloalkane
Greek
kyklos- circle, wheel, cycle; rotate
-alkyl- alcohol; a monovalent radical, such as ethyl or propyl
-ane a saturated hydrocarbon
An alicyclic hydrocarbon with a saturated ring; also called cycloparaffin.

Cyclonic

Greek

kyklos- circle, wheel, cycle; rotate

-ic (ikos) relating to or having some characteristic of
An atmospheric system characterized by the rapid inward circulation of air masses about a low-pressure center, usually accompanied by stormy, often destructive weather. Cyclones circulate counterclockwise in the Northern Hemisphere and clockwise in the Southern Hemisphere.

Cyclotron

Greek

kyklos- circle, wheel, cycle; rotate

-tron device for manipulating subatomic particles
A circular particle accelerator in which charged subatomic particles are accelerated outward in a plane perpendicular to a fixed magnetic field by an alternating electric field.

Cygnus

Latin

cygnus swan

A constellation in the Northern Hemisphere near Lacerta and Lyra, containing the star Deneb; also called the Northern Cross or the Swan.

Cystic

Greek

kustis- (cyst) sac or bladder that contains fluid

-ic (ikos) relating to or having some characteristic of
Of or related to a fluid-filled sac; a cyst or cystlike object. In anatomy, relating to the gallbladder or urinary bladder.

Cysticercus

Greek

kustis- (cyst) sac or bladder that contains fluid

-kerkos tail

A type of juvenile tapeworm in which an invaginated and introverted scolex is contained in a fluid-filled bladder.

Cystidolaparotomy

Greek

kustis- (cyst) sac or bladder that contains fluid

-lapar- soft part of the body between the ribs, hip, and flank; the loin

-tomos (temnein) to cut, incise, section
Incision of the bladder through the abdominal wall.

Cystitis

Latin

kustis- (cyst) sac or bladder that contains fluid

-itis inflammation
Inflammation of the urinary bladder.

Cystocele

Greek

kustis- (cyst) sac or bladder that contains fluid

-kele hernia, tumor
A herniation of the urinary bladder through the wall of the vagina.

Cystoscopy

Greek

kustis- (cyst) sac or bladder that contains fluid

-skopion for viewing with the eye
The process of examining the urinary bladder by looking into it with a scope instrument.

Cytoglucopenia

Greek

kutos- (cyto) sac or bladder that contains fluid

-gluc- glucose

-penia reduction, poverty, lack, deficiency
An intercellular deficiency of glucose.

Cytokine

Greek

kutos- (cyto) sac or bladder that contains fluid

-kinein to move

Any of several regulatory proteins, such as the interleukins and lymphokines, that are released by cells of the immune system and act as intercellular mediators in the generation of an immune response.

Cytokinesis

Greek

kutos- (cyto) sac or bladder that contains fluid

-kine- movement, motion

-sis action, process, state, condition
The division of the cytoplasm of a cell following the division of the nucleus.

Cytokinin

Greek

kutos- (cyto) sac or bladder that contains fluid

-kinein to move

Any of a class of plant hormones that promote cell division and growth and delay the senescence of leaves.

Cytolysis

Greek

kutos- (cyto) sac or bladder that contains fluid

-ly- (luein) to loosen, dissolve; dissolution, break

-sis action, process, state, condition
The dissolution or destruction of a cell.

Cytopharynx

Greek

kutos- (cyto) sac or bladder that contains fluid

-pharynx throat
Short tubular gullet in ciliate protozoa.

Cytoplasm

Greek

kutos- (cyto) sac or bladder that contains fluid
-plasm (plassein) to mold or form cells or tissues
Substance of the body of a cell excluding the nucleus.

Cytoproct

Greek

kutos- (cyto) sac or bladder that contains fluid
-proktos anus
Site on a protozoan where indigestible matter is expelled.

Cytopyge

Greek

kutos- (cyto) sac or bladder that contains fluid
-pyge rump, buttocks
In some protozoa, localized site for expulsion of waste.

Cytoskeleton

Greek

kutos- (cyto) sac or bladder that contains fluid
-skeletos dried body
A network of interconnected filaments and tubules that extends from the nucleus to the plasma membrane in eukaryotic cells.

Cytosol

Greek/Latin

kutos- (cyto) sac or bladder that contains fluid
-solvere to loosen
The fluid component of cytoplasm, excluding organelles and the insoluble, usually suspended cytoplasmic components.

Cytostome

Greek

kutos- (cyto) sac or bladder that contains fluid
-stoma mouth
The mouth of a unicellular organism, sometimes consisting of a hollow tube and a groovelike opening.

Cytotoxicity

Greek

kutos- (cyto) sac or bladder that contains fluid
-toxikos- poison
-ity state or quality of
The state or quality of being toxic to cells.

D

Dactylozooid
Greek
dactylo- finger, toe
-zoon- animal, animal-like
-oid (oeidēs) resembling, having the appearance of
A hydroid modified for catching prey; it is long, with tentacles or short knobs, and with or without a mouth.

Data
Latin
datum something given
Factual information, especially information organized for analysis or used to reason or make decisions.

Decantation
Latin
de- do or make the opposite of, reverse the action of, undo; from, apart, away
-canthus- rim of a wheel or vessel
-ion state, process, or quality of
The process of separating a mixture of two or more layers by pouring layers into separate containers.

Decapoda
Greek
deca- ten
-pod foot
The order of crustaceans, which includes the shrimps, lobsters, crabs, etc.

Decay
Latin
de- do or make the opposite of, reverse the action of, undo; from, apart, away
-cadere to fall
To break down into component parts.

Deciduous
Latin
decidu- to fall off
-ous full of, having the quality of, relating to
Falling off at a specific season or stage of growth.

Decipher
Latin/Arabic
de- do or make the opposite of, reverse the action of, undo; from, apart, away
-safira- to be empty
-er one that performs an action
To read, interpret, or convert complex, sometimes ambiguous data into a simplified form.

Declination
Greek
de- do or make the opposite of, reverse the action of, undo; from, apart, away
-klinein- to lean; sloping
-ation action, process, state, or condition
A measure of how far north or south an object is from the celestial equator.

Decomposer
Latin
de- do or make the opposite of, reverse the action of, undo; from, apart, away
-compose- to form, create
-er one that performs an action
Organism that feeds on and breaks down dead matter.

Defect
Latin
de- do or make the opposite of, reverse the action of, undo; from, apart, away
-fecere make, do, cause, produce, build

An imperfection that causes inadequacy or failure; a shortcoming.

Deglutination
Latin
de- do or make the opposite of, reverse the action of, undo; from, apart, away
-glutinare- to glue
-ion state, process, or quality of
The act of ungluing; the process of removing the gluten from flour.

Deglutition
Latin
de- do or make the opposite of, reverse the action of, undo; from, apart, away
-glūtīre- to gulp
-ion state, process, or quality of
The act or process of swallowing.

Degradation
Latin
de- do or make the opposite of, reverse the action of, undo; from, apart, away
-gradus- walk, step, take steps, move around; walking or stepping
-ion state, process, or quality of
To reduce the complexity of. In geology, the process of wearing away at the earth's surface through erosion.

Dehiscent
Latin
de- do or make the opposite of, reverse the action of, undo; from, apart, away
-hiare- to gape
-ent causing an action, being in a specific state; within
The opening of a fruit to liberate the seeds.

Deletion
Latin
deletus- to erase, destroy
-ion state, process, or quality of
The loss of a piece of chromosome that has broken away from the genetic material.

Deliquescent
Latin
deliquiscere melt by absorption of moisture
-ent causing an action, being in a specific state; within
A substance that absorbs enough water from the air that it dissolves completely to a liquid solution.

Dendrite
Greek
dendro- tree, resembling a tree
-ite a part of or product of
A branching, treelike extension from the body of a nerve cell that detects nerve impulses transmitted from the axons of other neurons.

Dendrochore
Greek
dendro- tree, resembling a tree
-chore a central and often foundational part, usually distinct from the enveloping part by a difference in nature
That part of the earth's surface covered by trees.

Dendrochronology
Greek
dendro- tree, resembling a tree
-khronos- time
-logy (logos) used in the names of sciences or bodies of knowledge
A method of dating using annual tree rings; tree ring chronology.

Dendroclastic
Greek
dendro- tree, resembling a tree
-klastos break, break in pieces
Breaking or destroying trees; a destroyer of trees.

Dendroclimatology
Greek
dendro- tree, resembling a tree
-klinein- to lean; sloping
-ate- characterized by having
-logy (logos) used in the names of sciences or bodies of knowledge
The determination of past climatic conditions from the study of the annual growth rings of trees.

Dendrohydrology
Greek
dendro- tree, resembling a tree
-hydr- water
-logy (logos) used in the names of sciences or bodies of knowledge
The study of tree ring configuration to determine hydrologic occurrences.

Density
Latin
densi- thick, thickly set, crowded, compact
-ity state of, quality of
The state or quality of being dense; compactness; closely set or crowded condition. Density is a measure of mass per unit of volume.

Dental
Latin
denti- teeth or tooth
-al of the kind of, pertaining to, having the form or character of
Of or relating to the teeth or to dentistry.

Dentalgia
Greek/Latin
denti- teeth or tooth
-algia pain, sense of pain; painful, hurting
An aching pain in or near a tooth; toothache.

Dentifrice
Latin
denti- teeth or tooth
-frice to rub; a rubbing
A powder or other preparation for cleansing or rubbing the teeth; a tooth powder or paste.

Dentition
Latin
denti- teeth or tooth
-ion state, process, or quality of
The number, type, and arrangement of an animal's teeth.

Deposit
Latin
de- do or make the opposite of, reverse the action of, undo; from, apart, away
-ponere to put
To lay down or leave behind by a natural process; to settle down in layers, as in mineral deposits.

Depressor
Latin
de- do or make the opposite of, reverse the action of, undo; from, apart, away
-premere- to press
-or a condition or property of things or persons; person who does something
A muscle that draws down a part of the body; a substance that slows a physiological activity.

Dermal
Greek
derm- skin
-al of the kind of, pertaining to, having the form or character of
Of or relating to the skin or dermis.

Dermatologist
Greek
dermat- skin
-logist one who deals with a specific topic
A physician who specializes in the diagnosis and treatment of skin disorders.

Dermatophyte
Greek
dermat- skin
-phyte plant
Any one of a number of fungi that infect the skin and nails.

Dermatozoon
Greek
dermat- skin
-zoon animal
Reference to animal skin or a branch of medicine dealing with animals.

Desiccator
Latin
desiccare make quite dry
A device used for drying substances; a closed glass vessel containing a deliquescent substance.

Desmoplastic
Greek
desmo- bond, adhesion
-plastos- (plassein) something molded; to mold
-ic (ikos) relating to or having some characteristic of
Pertaining to the production or formation of adhesions or fibrosis in the vascular connective tissue framework of an organ.

Detergent
Latin
de- out, off, apart, away
-terrēre- to frighten
-agere to do
A cleansing substance that acts similarly to soap but is made from chemical compounds rather than fats and lye.

Detritivore
Latin
deterere- to wear away, rub, grind; worn down
-vore eat, consume, ingest, devour
An organism that lives on dead and discarded organic matter; includes large scavengers, smaller animals such as earthworms and some insects, as well as decomposers (fungi and bacteria).

Detritus
Latin
deterere to lessen, wear away
Loose material (stone fragments, silt, etc.) that is worn away from rocks.

Deuterium
Greek
deuteros- second, two in number
-ium chemical element
An isotope of hydrogen with one proton and one neutron in the nucleus.

Deuterostome
Greek
deuteros- second, two in number
-stoma mouth
An animal whose mouth forms from an opening other than the blastopore.

Dextrorotatory

Latin
dextra- right or clockwise
-rota- wheel
-ory of or pertaining to
Rotating to the right in a plane of polarized light.

Diagnose

Greek
dia- through, across, apart
-gnose to know or learn
To arrive at a conclusion or determine the cause of a disorder or disease, usually by deductive reasoning.

Diagnosis

Greek
dia- through, across, apart
-gno- to come to know
-sis action, process, state, condition
The act or process of identifying or determining the nature and cause of a disease.

Diaheliotropism

Greek
dia- through, across, apart
-helio- sun
-trope- bend, curve, turn, a turning; response to a stimulus
-ism state or condition, quality
A tendency of leaves to have their dorsal surface toward the rays of the sun.

Dialysis

Greek
dia- through, across, apart
-ly- (*luein*) to loosen, dissolve; dissolution, break
-sis action, process, state, condition
The separation of smaller molecules from larger molecules or of dissolved substances from colloidal particles in a solution by selective diffusion through a semipermeable membrane.

Diamagnetic

Greek
dia- through, across, apart
-magnēs- stone from Magnesia (city in Asia Minor)
-ic (ikos) relating to or having some characteristic of
A substance that is weakly repelled by a magnet.

Diaphragm

Greek
dia- through, across, apart
-phragma fence
Muscular partition between the chest and abdominal cavities.

Diapsids

Greek
di- two

-apsis arch
Amniotes in which the skull bears two pairs of temporal openings; includes reptiles (except turtles) and birds.

Diarrhea

Greek
dia- through, across, apart
-rhein to flow or run
Frequent and possibly excessive elimination of watery feces.

Diastereomer

Greek
di- two
-a- without, not
stereos- being of three dimensions
-mer one that has
Two compounds that are optical isomers that are not mirror images of each other, with different physical properties and reactivity.

Diastole

Greek
diast- dilation, spreading
-ole little
Relaxation period of a heart during the cardiac cycle.

Diatom

Greek
dia- through, across, apart
-tomos (temnein) to cut, incise, section
Any of a class of minute planktonic unicellular or colonial algae with silicified skeletons that form diatomite.

Diatomic

Latin
di- two, twice, double
-a- no, absence of, without, lack of, not
-tomos- (temnein) to cut, incise, section
-ic (ikos) relating to or having some characteristic of
Consisting of or relating to a molecule that is composed of two atoms.

Dichotomy

Greek
dicho- akin to
-tomos (temnein) to cut, incise, section
A dividing or branching into two equal parts.

Dichroism

Greek
di- two, twice, double
-khrōma- color
-ism state or condition, quality
The property of showing two different colors at different concentrations or when viewed at different angles.

Dicotyledon
Greek
di- two, twice, double
-kotuledon a kind of plant; a seed leaf; a hollow
or cup-shaped object
Flowering plant group whose members have two
embryonic leaves.

Dictyostele
Greek
dictyo- net, netlike
-stele pillar
In some ferns, a stele that is interrupted by leaf
gaps so as to resemble a network of strands.

Diencephalon
Greek
dia- through
-enkephalos in the head
The posterior portion of the forebrain; includes
areas of the midbrain such as the thalamus and
hypothalamus.

Differentiation
Latin
differre- to differ; delay
-atus- in
-ion state, process, or quality of
The process by which cells or tissues undergo a
change toward a more specialized form or function, especially during embryonic development.

Diffraction
Latin
dis- undo; apart, in all directions
-frangere- to break
-ion state, process, or quality of
Change in the directions and intensities of a
group of waves after passing by an obstacle or
through an aperture whose size is approximately
the same as the wavelength of the waves.

Diffusion
Latin
diffundere- to spread out
-ion state, process, or quality of
The process in which particles in a fluid move
from an area of higher concentration to an area of
lower concentration.

Digest
Latin
digerere to break down
To break into smaller parts and simpler compounds.

Digestion
Latin
di- apart, away, from
-gerere- to bear

-ion state, process, or quality of
The ability to change into absorbable form.

Digitigrade
Latin
digitus- finger or toe
-gradus step or degree
Walking on the digits with the posterior part of
the foot raised.

Dihybrid
Greek
di- two, twice, double
-hybrida- mongrel offspring
-id state, condition; having, being, pertaining to,
tending to, inclined to
The offspring of parents differing in two specific
gene pairs.

Dilation
Latin
di- apart, away, from
-lātus wide
The process of becoming wider or larger, as of a
blood vessel.

Dilute
Latin
di- apart, away, from
-luere wash, clean
To make thinner or less concentrated by adding a
liquid such as water.

Dimension
Latin
dis- undo; apart, in all directions
-metiri- to measure out
-ion state, process, or quality of
A measurement of spatial extent; specifically, one of
three coordinates determining a position in space.

Dimorphism
Greek
di- two, twice, double
-morph- shape, form, figure, or appearance
-ism state or condition
The existence within a species of two distinct
forms according to color, sex, organ structure, or
other characteristic.

Dinoflagellate
Greek
dinos- whirling
-flagrum- whip
-ate characterized by having
A marine protozoan of the order Dinoflagellata,
having two flagella and a cellulose covering and
forming one of the chief constituents of plankton.

They include bioluminescent forms and forms that produce red tide.

Dinosaur
Greek
deinos- terrible, monstrous
-sauros lizard
A variety of extinct reptiles that existed during the Mesozoic era.

Dioecious
Greek
di- two, twice, double
-oec- environment, habitat
-ious full of, having the quality of, relating to
Having the male and female reproductive organs in separate individuals.

Diphycercal
Greek
diphues- twofold
-kerkos- tail
-al of the kind of, pertaining to, having the form or character of
Having a tail that tapers to a point, as in lungfishes; the vertebral column extends to tip without upturning.

Diphyodont
Greek
di- two, twice, double
-phuein- to grow
-odont having teeth
Having deciduous and permanent sets of teeth successively.

Diploblastic
Greek
diploos- double
-blastos bud, germ cell
-ic (ikos) relating to or having some characteristic of
Referring to an organism with two germ layers, endoderm and ectoderm.

Diploid
Greek
diploos- double
-oid (oeidēs) resembling, having the appearance of
Having the somatic (double, or $2n$) number of chromosomes, or twice the number characteristic of a gamete of a given species.

Diplopia
New Latin
diploos- double
-optic- eye, optic
-ia names of diseases, place names, or Latinizing plurals

Condition in which two images of a single object are seen due to unequal action of the eye muscles; also called double vision.

Dipole
Middle English from Old French (from Latin)
di- two, twice, double
-pole either of two oppositely charged terminals
A pair of equal and opposite electrical charges or magnetic poles, separated by a small distance.

Disaccharide
Greek
di- two, twice, double
-saccharon- sugar
-ide group of related chemical compounds
Any class of sugars, including lactose and sucrose, that are composed of two monosaccharides; a double sugar.

Disease
Middle French
dis- apart, away from; utterly, completely, in all directions
-aise ease, freedom from pain
A condition of the living animal or plant body or of one of its parts that impairs normal functioning.

Dispersion
Latin
dis- apart, away from; utterly, completely, in all directions
-spargere- to scatter or strew; sprinkle
-ion state, process, or quality of
The passing out or spreading about of something.

Dispersoid
Latin
dis- apart, away from; utterly, completely, in all directions
-spargere- to scatter or strew; sprinkle
-oid (oeidēs) resembling, having the appearance of
A substance consisting of finely divided particles dispersed in a medium.

Displacement
Greek
dis- apart, away fro;, utterly, completely, in all directions
-place- to put in or as if in a particular place or position
-ment state or condition resulting from a (specified) action
A vector or the magnitude of a vector from an initial position to a subsequent position assumed by a body.

Dissection

Latin
dis- apart, away from; utterly, completely, in all directions
-sectus- to cut
-ion state, process, or quality of
The separation of a whole into its parts for study.

Disseminate

Latin
dis- apart, away from; utterly, completely, in all directions
-seminare- to plant or propagate (from *semen, seminis,* meaning "seed")
-ate characterized by having
To scatter for growth and propagation; to spread, to diffuse.

The Black Death

The black plague struck continental Europe in the year 1347. Without a doubt, it was one of the most devastating natural disasters ever to befall humankind. In many ways it altered the course of human history. The epidemiology of plague was a mystery to all. Even while it was happening, no one really knew its cause, let alone its cure. Thousands of people died, and others fled. Those who treated the very ill died. Those who buried the dead died.

Today, historians and scientists believe that the Black Death stemmed from a microorganism called *Yersinia pestis,* a bacterium that was carried and spread by fleas living on black rats. During that era, the black rat population vastly exceeded that of the larger and fiercer Norwegian gray rat. Interestingly, the Norwegian gray rat was a poor vector for the fleas carrying the bacteria.

In the late 1370s and early 1380s, Marchione di Coppo Stefani wrote the descriptive narrative *The Florentine Chronicle on Medieval Plague.* Excerpts from that essay describe the horror of the plague:

In the year of the Lord 1348 there was a very great pestilence in the city and district of Florence. It was of such a fury and so tempestuous that in houses in which it took hold previously healthy servants who took care of the ill died of the same illness. Almost none of the ill survived past the fourth day. Neither physicians nor medicines were effective. Whether because these illnesses were previously unknown or because physicians had not previously studied them, there seemed to be no cure. There was such a fear that no one seemed to know what to do. When it took hold in a house it often happened that no one remained who had not died. And it was not just that men and women died, but even sentient animals died. Dogs, cats, chickens, oxen, donkeys, sheep showed the same symptoms and died of the same disease. And almost none, or very few, who showed these symptoms, were cured. The symptoms were the following: a bubo in the groin, where the thigh meets the trunk; or a small swelling under the armpit; sudden fever; spitting blood and saliva (and no one who spit blood survived it). It was such a frightful thing that when it got into a house, as was said, no one remained. Frightened people abandoned the house and fled to another. Those in town fled to villages. Physicians could not be found because they had died like the others. And those who could be found wanted vast sums in hand before they entered the house. And when they did enter, they checked the pulse with face turned away. They inspected the urine from a distance and with something odoriferous under their nose. Child abandoned the father, husband the wife, wife the husband, one brother the other, one sister the other. In all the city there was nothing to do but to carry the dead to a burial. And those who died had neither confessor nor other sacraments. And many died with no one looking after them. And many died of hunger because when someone took to bed sick, another in the house, terrified, said to him:

"I'm going for the doctor." Calmly walking out the door, the other left and did not return again. Abandoned by people, without food, but accompanied by fever, they weakened. There were many who pleaded with their relatives not to abandon them when night fell. But [the relatives] said to the sick person, "So that during the night you did not have to awaken those who serve you and who work hard day and night, take some sweetmeats, wine or water. They are here on the bedstead by your head; here are some blankets." And when the sick person had fallen asleep, they left and did not return. If it happened that he was strengthened by the food during the night he might be alive and strong enough to get to the window. If the street was not a major one, he might stand there a half hour before anyone came by. And if someone did pass by, and if he was strong enough that he could be heard when he called out to them, sometimes there might be a response and sometimes not, but there was no help. No one, or few, wished to enter a house where anyone

(Continued)

was sick, nor did they even want to deal with those healthy people who came out of a sick person's house. And they said to them: "He is stupefied, do not speak to him!" saying further: "He has it because there is a bubo in his house." They call the swelling a bubo. Many died unseen. So they remained in their beds until they stank. And the neighbors, if there were any, having smelled the stench, placed them in a shroud and sent them for burial. The house remained open and yet there was no one daring enough to touch anything because it seemed that things remained poisoned and that whoever used them picked up the illness. At every church, or at most of them, they dug deep trenches, down to the waterline, wide and deep, depending on how large the parish was. And those who were responsible for the dead carried them on their backs in the night in which they died and threw them into the ditch, or else they paid a high price to those who would do it for them. The next morning, if there were many [bodies] in the trench, they covered them over with dirt. And then more bodies were put on top of them, with a little more dirt over those; they put layer on layer just like one puts layers of cheese in a lasagna.

The beccamorti [literally, vultures] who provided their service, were paid such a high price that many were enriched by it. Many died from [carrying away the dead], some rich, some after earning just a little, but high prices continued. Servants, or those who took care of the ill, charged from one to three florins per day and the cost of things grew. The things that the sick ate, sweetmeats and sugar, seemed priceless. Sugar cost from three to eight florins per pound. And other confections cost similarly. Capons and other poultry were very expensive and eggs cost between twelve and twenty-four pence each; and he was blessed who could find three per day even if he searched the entire city. Finding wax was miraculous. A pound of wax would have gone up more than a florin if there had not been a stop put [by the communal government] to the vain ostentation that the Florentines always make [over funerals]. Thus it was ordered that no more than two large candles could be carried [in any funeral]. Churches had no more than a single bier which usually was not sufficient. Spice dealers and beccamorti sold biers, burial palls, and cushions at very high prices. Dressing in expensive woolen cloth as is customary in [mourning] the dead, that is in a

long cloak, with mantle and veil that used to cost women three florins climbed in price to thirty florins and would have climbed to 100 florins had the custom of dressing in expensive cloth not been changed. The rich dressed in modest woolens, those not rich sewed [clothes] in linen. Benches on which the dead were placed cost like the heavens and still the benches were only a hundredth of those needed. Priests were not able to ring bells as they would have liked. Concerning that [the government] issued ordinances discouraging the sounding of bells, sale of burial benches, and limiting expenses. They could not sound bells, sell benches, nor cry out announcements because the sick hated to hear of this and it discouraged the healthy as well. Priests and friars went [to serve] the rich in great multitudes and they were paid such high prices that they all got rich. And therefore [the authorities] ordered that one could not have more than a prescribed number [of clerics] of the local parish church. And the prescribed number of friars was six. All fruits with a nut at the center, like unripe plums and unhusked almonds, fresh broadbeans, figs and every useless and unhealthy fruit, were forbidden entrance into the city. Many processions, including those with relics and the painted tablet of Santa Maria Inpruneta, went through the city crying our "Mercy" and praying and then they came to a stop in the piazza of the Priors. There they made peace concerning important controversies, injuries and deaths. This [pestilence] was a matter of such great discouragement and fear that men gathered together in order to take some comfort in dining together. And each evening one of them provided dinner to ten companions and the next evening they planned to eat with one of the others. And sometimes if they planned to eat with a certain one he had no meal prepared because he was sick. Or if the host had made dinner for the ten, two or three were missing. Some fled to villas, others to villages in order to get a change of air. Where there had been no [pestilence], there they carried it; if it was already there, they caused it to increase. None of the guilds in Florence was working. All the shops were shut, taverns closed; only the apothecaries and the churches remained open. If you went outside, you found almost no one. And many good and rich men were carried from home to church on a pall by four beccamorti and one tonsured clerk who carried the cross. Each of them wanted a florin.

(Continued)

This mortality enriched apothecaries, doctors, poultry vendors, beccamorti, and greengrocers who sold of poultices of mallow, nettles, mercury and other herbs necessary to draw off the infirmity. And it was those who made these poultices who made a lot of money. Woolworkers and vendors of remnants of cloth who found themselves in possession of cloths [after the death of the entrepreneur for whom they were working] sold it to whoever asked for it. When the mortality ended, those who found themselves with cloth of any kind or with raw materials for making cloth was enriched. But many [who actually owned cloths being processed by workers] found it to be moth-eaten, ruined or lost by the weavers. Large quantities of raw and processed wool were lost throughout the city and countryside.

This pestilence began in March, as was said, and ended in September 1348. And people began to return to look after their houses and possessions. And there were so many houses full of goods without a master that it was stupefying. Then those who would inherit these goods began to appear. And such it was that those who had nothing found themselves rich with what did not seem to be theirs and they were unseemly because of it. Women and men began to dress ostentatiously.

Dissociation
Latin
dis- apart, away from; utterly, completely, in all directions
-sociar- to join
-ion state, process, or quality of
The process by which a chemical combination breaks up into simpler constituents.

Distillation
Latin
dis- apart, away from; utterly, completely, in all directions
-stillare- to drip or trickle
-ion state, process, or quality of
A process used to separate a liquid mixture based on the boiling points of the substances within the solution.

Distribution
Latin
dis- apart, away from; utterly, completely, in all directions
-tribuere- to give
-ion state, process, or quality of
In mathematics, sample values presented in order from the lowest to the highest.

Diurnal
Latin
diurnus- day
-al of the kind of, pertaining to, having the form or character of
Related to or occurring within a twenty-four-hour period; occurring in the daytime hours rather than the nighttime hours.

Diverge
Latin
di- two, twice, double
-verge to tend to move in a particular direction
To go or extend in different directions from a common point.

Diverticulum
Latin
de- reverse the action of, undo; from, apart, away
-vertere- to turn
-um (**singular**) structure
-a (**plural**) structure
A pouchlike structure extending out or away from an organ such as the intestines.

DNA ligase
Latin
ligo- bind, tie
-ase enzyme
Enzyme that links DNA fragments; used during the production of recombinant DNA to join foreign DNA to the vector DNA.

Dodecahedron
Greek
dodeca- twelve
-hedron face
A Platonic solid with twelve faces; the fifth essence.

Doldrums
Middle English
dold to dull
-um (**singular**) structure
-a (**plural**) structure
A region of the ocean near the equator, characterized by calms, light winds, and squalls.

Domain
Latin
dominus lord
Any of numerous contiguous regions in a ferromagnetic material in which the direction of spontaneous magnetization is uniform and different from that in neighboring regions.

Dominant (traits)

Latin

dominan dominant

The hereditary traits that exhibit a stronger influence on the phenotype than their more recessive alleles.

Doping

Dutch

doopen- to dip

-ing the act of or action

The act of introducing impurities into a crystal structure in order to acquire useful properties.

Dormant

Latin

dormire- to sleep

-ant a person who, the thing which

Describes an inactive state of a seed.

Dorsal

Latin

dorsalis- back

-al of the kind of, pertaining to, having the form or character of

Of, toward, on, in, or near the back or upper surface of an organ, part, or organism.

Downburst

Swedish

dun- down

-bresta to break asunder

Violent downdrafts that are concentrated in a local area.

Drag

Old Norse

draga to draw, drag

The retarding force exerted on a moving body by a fluid medium such as air or water.

Drosophila

Greek

drosos– dew

-philos beloved

Any of various small fruit flies of the genus *Drosophila*.

Drought

Anglo-Saxon

dygre dry

Dryness; lack of rain or water.

Drumlin

Scottish Gaelic

drum- ridge, back; long, narrow hill

-lin small or little

An elongated hill or ridge of glacial drift; elongated landform that results when a glacier moves over an older moraine.

Pythagoras of Samos

During the reign of the tyrant Polycrates (535–515 BC), the Greek island of Samos in the eastern Aegean Sea was home to Pythagoras. He was one of the most influential mathematicians and philosophers of his time. All those who truly appreciate mathematics hold a special place in their hearts for the Pythagoreans, who believed that numbers constitute the true nature and harmony of the world—indeed, the universe. That is, the synchronization of the universe relies on mathematical harmony. The Pythagoreans did not believe in experimentation. They relied on the faculties of thought, reason, and deduction. Pythagoras' followers (who called themselves the *mathematikoi*) reasoned that the relationships among all things were mathematical. Even the workings of the mind (logic and reason) were, to the Pythagoreans, the result of mathematical expressions.

Pythagoras is given credit for developing a mathematical correlation between whole numbers and musical scales. He and his followers are recognized for developing the Pythagorean theorem, which is well known among all who study geometry. Beauty was to be found in the shapes of solids. The four regular solids, the tetrahedron, hexahedron (cube), octahedron, and icosahedron, represented the four elements (earth, fire, air, and water), the "roots" of the earth. There was a mystical, almost fearful forbiddance directed toward the fifth of the regular solids, the dodecahedron. The Pythagoreans believed that the twelve pentagons that form the sides of this solid were somehow celestial and not of this earth. This fifth element, which could only come from the heavens, signified by the dodecahedron gave rise to the term *quintessence*: the purest, most highly concentrated essence, the "fifth essence."

Ductile

Latin

ductus- to be hammered out into a tube or pipe; leading or drawing

-ile changing; ability; suitable; tending to

Property of a metal that enables it to be easily drawn into a wire.

Dunite

English
dun- referring to Mount Dun in New Zealand
-ite minerals and fossils
A dense igneous rock that consists mainly of olivine and is a source of magnesium.

Duodenostomy

Latin/Greek
duodecum- twelve
-stoma- opening
-y place for an activity; condition, state
The surgical establishment of an opening into the duodenum.

Duodenum

Latin
duodeni- twelve each
-um (**singular**) structure
-a (**plural**) structure
The beginning portion of the small intestine, approximately 12 inches in length, starting at the lower end of the stomach and extending to the jejunum.

Duramen

Latin/Middle English
durare- to harden; hard growth
-enen to cause or become
The older, nonliving central wood of a tree or woody plant, usually darker and harder than the younger sapwood.

Dynamic

Greek
dunamikos- powerful
-ic (ikos) relating to or having some characteristic of
Marked by usually continuous and productive activity or change; of or relating to energy or to objects in motion.

Dysentery (amoebic)

Greek
dys- painful, difficult, disordered, impaired, defective, ill
-enteron- intestines
-y place for an activity, condition, state
Extreme diarrhea with blood in the feces, caused by either the ingestion of certain bacteria (shigella) or protozoa (*Entamoeba hystolitica*).

Dysfunction

Greek/Latin
dys- painful, difficult, disordered, impaired, defective, ill
-fungi- performance, execution
-ion state, process, or quality of
Abnormal, inadequate, or impaired function of an organ or body part.

Dyslexia

Greek
dys- painful, difficult, disordered, impaired, defective, ill
-legein- word, speech
-al of the kind of, pertaining to, having the form or character of
A disorder affecting the comprehension and use of words.

Dyspepsia

Greek
dys- painful, difficult, disordered, impaired, defective, ill
-peps- digestion
-ia names of diseases, place names, or Latinizing plurals
Commonly referred to as indigestion, a painful disorder of the stomach.

Dysphagia

Greek
dys- painful, difficult, disordered, impaired, defective, ill
-phage- to eat
-ia names of diseases, place names, or Latinizing plurals
Difficulty in swallowing, but not to be confused with painful swallowing. Dysphagia is a symptom of numerous paralytic diseases, including amyotrophic lateral sclerosis (Lou Gehrig's disease).

Dyspnea

Greek
dys- painful, difficult, disordered, impaired, defective, ill
-pnoia breathing or breath
Sensation of difficult or labored breathing.

Dystrophy

Greek
dys- painful, difficult, disordered, impaired, defective, ill
-trophos- (trophein) to nourish; food, nutrition; development
-y place for an activity; condition, state
Any of several disorders involving atrophy of muscular tissue.

E

Eccentric
Greek
ek- out of
-kentron- center
-ic (ikos) relating to or having some characteristic of
Deviating from a circular form or path, as an elliptical orbit.

Eccentricity
Greek
ek- out of
-kentron- center
-itas variant
The measure of the degree of elongation of an ellipse. For example, a circle has an eccentricity of 0, and a parabola (an open figure) has an eccentricity of 1.

Eccrine
Greek
ek- out of
-krinein to separate
Applies to a type of mammalian sweat gland that produces a watery secretion.

Ecdysiotropin
Greek
ekdysis- to strip off; escape
-trope- bend, curve, turn, a turning; response to a stimulus
-in protein or derived from a protein
Hormone secreted in the brain of insects that stimulates the prothoracic gland to secrete molting hormone.

Ecdysone
Greek
ekdusis- to shed or molt
-one a chemical compound containing oxygen in a carbonyl group
A steroid hormone, produced by the prothoracic gland of insects, that promotes growth and controls molting.

Echinoderma
Greek
echino- spiny, hedgehog
-derma skin
Radially symmetrical marine invertebrates, including starfish and sea urchins.

Echocardiograph
Greek
ēkhō- repeat of sound
-kard- heart, pertaining to the heart
-graphia (graphein) to write, record, draw, describe
A technological instrument designed to noninvasively transmit ultrasonic impulses into the chest that are reflected back so that the heart can be imaged and studied.

Echolocation
Greek
ēkhō- repeat of sound
-locare to place
A sensory adaptation used by certain animals such as dolphins and bats. Pulses of sound waves are emitted by the animal and reflected back from an object; the organism can then determine the distance of the object by the elapsed time.

Eclipse
Greek
ektos- outer, external, out of, out, outside; away from
-leipein to leave
The partial or complete obscuring, relative to a designated observer, of one celestial body by another.

Ecliptic
Greek
ektos- outer, external, out of, out, outside; away from
-lipo- abandon, to leave [behind]
-ic (ikos) relating to or having some characteristic of
The apparent path of the sun traced along the sky in the course of the year.

Ecocentrism
Greek
oikos- home, house
-centr- center
-ism state or condition
The view or belief that environmental concerns should take precedence over the needs and rights of human beings.

Ecocide
Greek
oikos- home, house
-cide (caedere) to cut, kill, hack at, or strike
Destruction or damage to the environment, especially intentionally (e.g., by herbicides in war).

Ecogenetics
Greek/Latin
oikos- home, house
-gen- to give birth, kind, produce
-ic (ikos) relating to or having some characteristic of
The study of the relationship between genetic factors and the nature of response to an environmental agent.

Ecohazard
Greek/Arabic
oikos- home, house
-az zahr the gaming die, dice game
Any activity or substance that may constitute a threat to a habitat or environment.

Ecology
Greek
oikos- house
-logy (logos) used in the names of sciences or bodies of knowledge
The science of the relationships between organisms and their environments

Ecosystem
Greek/Latin
oikos- home, house
-systema the universe.

The Eclipse That Stopped a War

Thales of Miletus (ca. 635–543 BC) is regarded by many as the father of science. He was a philosopher and an astronomer living in a time before Socrates. Unlike most philosophers of this time, he put his intellect to use in matters other than pure philosophy. Although his motive probably was not to become wealthy, he proved that by applying what he had learned about the natural world, he could succeed in business and politics. And he did. He was numbered among the Seven Sages of Greece, those statesmen who were known for their practical wisdom.

Thales studied the natural world and its events. He believed that the world was not created by supernatural forces, but rather by naturally occurring events. It was recorded by the historian Herodotus of Halicarnassus (ca. 484–425 BC) that Thales predicted the occurrence of a total solar eclipse on May 28, 585 BC. As it happened, that eclipse ended a long and bloody war. The warring factions, the Lydians and the Medes, were in the sixth year of a struggle with no end in sight. Right in the middle of the battle of Halys, "the day was turned into night," and the battle was stopped and the war ended.

An ecological community together with its environment, functioning as a unit.

Ecotaxis
Greek
oikos- home, house
-taxi arrangement, order; to put in order
The "homing" of recirculating lymphocytes to specific compartments of peripheral lymphoid tissues, with B cells going to B-dependent areas and T cells to T-dependent areas.

Ecotone
Latin
oikos- home, house
-tonos tension, pressure
A transition region where adjacent biomes blend, containing some organisms from each of the adjacent biomes plus some that are characteristic of, and perhaps restricted to, the ecotone; this region tends to have more species and to be more densely populated than either adjacent biome.

Ecotoxicologist

Greek

oikos- home, house

-toxikos- poison

-ologist one who deals with a specific topic

A specialist in the harmful effects of chemicals on the natural environment.

Ectobiology

Greek

ektos- outside, external, beyond

-bios- life, living organisms or tissue

-logy (logos) used in the names of sciences or bodies of knowledge

The study of the properties and biochemical constitution of the cell surface and the specific enzymes at the surface.

Ectocardia

Greek

ektos- outside, external, beyond

-kard- heart, pertaining to the heart

-ia names of diseases, place names, or Latinizing plurals

The congenital displacement of the heart, either inside or outside the thorax.

Ectoderm

Greek

ektos- outside, external, beyond

-derm skin

Embryonic tissue layer that leads to the differentiation of epidermal, nervous, and sensory organs and tissues.

Ectognatous

Greek

ektos- outside, external, beyond

-gnathos jaw

Derived characteristic of most insects, in which mandibles and maxillae are not in pouches.

Ectohormone

Greek

ektos- outside, external, beyond

-hormo- to rouse or to set in motion

-one chemical compound containing oxygen in a carbonyl group

A parahormonal chemical mediator of ecological significance that is secreted, largely by an organism (usually an invertebrate) into its immediate environment (air or water); it can alter the behavior or functional activity of a second organism, often of the same species as that secreting the ectohormone.

Ectolecithal

Greek

ektos- outside, external, beyond

-lekithos egg yolk

Yolk for nutrition of the embryo contributed by cells that are separate from the egg cell and are combined with the zygote by envelopment within the eggshell.

Ectomorphic

Greek

ektos- outside, external, beyond

-morph- shape, form, figure, or appearance

-ic (ikos) relating to or having some characteristic of

Referring to an individual characterized by having a lean, slightly muscular build in which tissues derived from the embryonic ectoderm predominate.

Ectoplasm

Greek

ektos- outside, external, beyond

-plasm (plassein) to mold or form cells or tissues

The cortex of a cell or that part of cytoplasm just under the cell surface.

Ectoscopy

Greek

ektos- outside, external, beyond

-skopein- see, view, sight, look at, examine

-y place for an activity; condition, state

A diagnostic method based on observation of chest and abdominal movements and said to be capable of determining the outlines of the lungs and of localized internal conditions.

Ectothermic

Greek

ektos- outside, external, beyond

-thermos- combining form of "hot" (heat)

-ic (ikos) relating to or having some characteristic of

Having a body temperature derived by heat acquired from the environment.

Edema

Greek

oidēma a swelling

The accumulation of excessive amounts of serous fluids in the tissues or cavities within the body.

Effect

Latin

ex- outside, outward, out of, out; away from

-facere- to do; carry, bear, bring

The result or consequence of an action.

Effector

Latin

ex- outside, outward, out of, out; away from

-facere- to do; carry, bear, bring

-or a condition or property of things or persons, person who does something

An organ or structure that responds as a result of nervous stimulation.

Efferent

Latin

ex- outside, outward, out of, out; away from
-facere- to do; carry, bear, bring
-ent causing an action; being in a specific state; within

Leading or conveying away from some organ— for example, nerve impulses conducted away from the brain, or blood conveyed away from an organ; contrasts with *afferent*.

Efficiency

Latin

efficere- to effect
-cy state, condition, quality

The ratio of useful work accomplished by a machine compared to the total work put into it; usually expressed as a percentage.

Effloresce

Latin

ex- outside, outward, out of, out; away from
-florere flower; to blossom

To become covered by a crusty deposit when water evaporates.

Ejecta

Latin

eicere- to throw out

Ejected matter, such as that from an erupting volcano.

Ejection

Latin

eicere- to throw out
-ion state, process, or quality of

The act of ejecting or the condition of being ejected.

Elastic

Greek

elaunein- to beat out
-ic (ikos) relating to or having some characteristic of

Returning to or capable of returning to an initial form or state after deformation.

Electricity

Greek

ēlektron- charge, electricity; dealing with positive and negative charges
-ity state or quality

The flow of electrons in a circuit. The speed of electricity is the speed of light (approximately 186,000 miles per second). In a wire, it is slowed due to the resistance in the material.

Electrocardiograph

Greek

ēlektron- charge, electricity; dealing with positive and negative charges
-kard- heart, pertaining to the heart
-graphia (graphein) to write, record, draw, describe

An instrument for recording the potential of the electrical currents that traverse the heart and initiate its contraction.

Electrodialysis

Greek

ēlektron- charge, electricity; dealing with positive and negative charges
-dia- through, across, point to point
-ly- loosening, dissolving, dissolution, breaking
-sis action, process, state, condition

A form of dialysis in which the application of current to electrodes is used to separate substances or compounds. Salt is removed from seawater in large quantities in this manner.

Electrolysis

Greek

ēlektron- charge, electricity; dealing with positive and negative charges
-ly- (luein) to loosen, dissolve; dissolution, break
-sis action, process, state, condition

A process in which electrolytes are created by splitting compounds using electric current.

Electrolyte

Latin/Greek

ēlektron- charge, electricity; dealing with positive and negative charges
-lyte substance capable of undergoing decomposition

A substance that when dissolved in a suitable solvent becomes an ionic conductor.

Electromagnetic

Greek

ēlektron- charge, electricity; dealing with positive and negative charges
-magnes- something that attracts (figurative sense)
-ic (ikos) relating to or having some characteristic of

Variation in electric and magnetic fields taking place in regular, repeating fashion.

Electron

Greek

ēlektron- charge, electricity; dealing with positive and negative charges
-on a particle

An elementary particle consisting of a charge of negative electricity equal to about 1.602×10^{-19} coulomb and having a mass when at rest of about 9.109534×10^{-28} gram, or about 1/1836 that of a proton.

Electronegativity

Greek

ēlektron- charge, electricity; dealing with positive and negative charges

-negare- say no, deny
-ity state or quality
Property of an element that indicates how strongly its atom attracts electrons in a chemical bond.

Electrophile
English
ēlektron- charge, electricity; dealing with positive and negative charges
-phile one who loves or has a strong affinity or preference for
A chemical compound or group attracted to electrons and tending to accept them.

Electrophoresis
Greek
ēlektron- charge, electricity; dealing with positive and negative charges
-phoros- being carried, bearing
-sis action, process, state, condition
The movement of suspended particles in a fluid under the influence of an electric field.

Electroweak
Greek/Middle English
ēlektron- charge, electricity; dealing with positive and negative charges
-weike pliant
Of or relating to the combination of the electromagnetic and weak nuclear forces in a unified theory.

Element
Latin
elementum rudiment, first principle
A substance that cannot be separated into simpler substances by chemical means.

Elimination
Latin
eliminat- to banish
-ion state, process, or quality of
A process by which wastes are removed from the body.

Ellipse
Latin/Greek
en- in, at, onto
-leipein to leave
A plane curve, especially a conic section whose plane is not parallel to the axis, base, or generatrix of the intersected cone.

Elliptical
Greek
elleiptikos- of a leaf shape; in the form of an ellipse
-al of the kind of, pertaining to, having the form or character of
Of, relating to, or having the shape of an ellipse; containing or characterized by ellipsis.

Elongation
Latin
elongate- to make or grow longer
-ion state, process, or quality of
The act of making something longer or the condition of being made longer.

Elytra
Greek
elutron sheath
The thickened or leathery forewings of insects such as beetles.

Embolism
Greek
em- in
-bol- (ballein) to put or throw
-ism state or condition
Obstruction or occlusion of a blood vessel blocking the flow of blood.

Embryo
Greek
em- in
-bruein to be full, bursting
An organism in its early stage of development, especially before it has reached a distinctively recognizable form.

Embryogenesis
Greek
em- in
-bruein- to be full, bursting
-gen- to give birth, kind, produce
-sis action, process, state, condition
The origin and development of the embryo; embryogeny.

Emigration
Latin
e- out
-migrare- to move
-ion state, process, or quality of
The act or process of leaving an area or country to live in another country.

Emission
Latin
ēmittere- to send out
-ion state, process or quality of
A substance discharged into the air, especially by an internal combustion engine.

Emphysema
Greek
em- in, into, inward; within
-phusan to blow
A pathological condition of the lungs marked by an abnormal increase in the size of the air spaces,

resulting in labored breathing and an increased susceptibility to infection.

Empirical
Greek
empeirikos- doctor relying on experience alone
-al of the kind of, pertaining to, having the form or character of
Referring to a formula that gives the simplest whole number ratio of atoms of elements in a compound.

Emulsification
Greek
-mulgēre- to milk out
-ation action, process, state, or condition
Process of mixing two liquids that do not dissolve in each other.

Emulsify
Latin
-mulgēre- to milk out
-fy cause; to become, make
To make into an emulsion.

Emulsion
Latin
ex- outside, outward, out of, out; away from
-mulgēre- to milk out
-ion state, process, or quality of
A suspension of small globules of one liquid in a second liquid with which the first will not mix.

Enantiomer
Greek
en- to cause to be
-anti- opposite
-mere considered apart from anything else; pure
Either of a pair of crystals, molecules, or compounds that are mirror images but not identical.

Encephalitis
Greek
en- in, into, inward; within
-cephalo- (kephalikos) head
-itis inflammation, burning sensation
Inflammation of the brain, usually caused by a viral infection.

Encephalomalacia
Greek
en- in, into, inward; within
-cephalo- (kephalikos) head
-malacia softening of tissue
Softening of brain tissue, usually caused by vascular insufficiency or degenerative changes.

Endemic
Greek
en- in, into, inward; within
-demo- population
-ic (ikos) relating to or having some characteristic of
A condition, such as a disease, that is prevalent in a specific area.

Endergonic
Greek
endo- inside, within
-ergon- work
-ic (ikos) relating to or having some characteristic of
A chemical reaction requiring energy to obtain the end products.

Endoabdominal
Greek
endo- inside, within
-abdomen- belly, venter, abdomen
-al of the kind of, pertaining to, having the form or character of
Relating to tissues and other materials found within the abdominal walls.

Endobenthos
Greek
endo- inside, within
-benthos deep; the fauna and flora of the bottom of the sea
Organisms living within the sediment on the seabed or lake floor.

Endocrine
Greek
endo- within
-krinein to separate
Glands that secrete hormones into the blood.

Endocytosis
Greek
endo- inside, within
-kutos- (cyto) sac or bladder that contains fluid
-sis action, process, state, condition
The process of moving things to the inside of a cell.

Endoderm
Latin
endo- inside, within
-derma skin
In animals, the inner layer of embryonic tissue from which the digestive organs develop.

Endoergic
Greek
endo- inside, within
-ergon- work
-ic (ikos) relating to or having some characteristic of
Occurring with absorption of energy. In biology, the process by which heat is generated to maintain a constant body temperature.

Endognathous
Greek
endo- inside, within
-gnathos jaw
Ancestral character of insects, found in the orders Diplura, Collembola, and Protura, in which the mandibles and maxillae are located in pouches.

Endolecithal
Greek
endo- inside, within
-ekithos yolk
Yolk for nutrition of the embryo incorporated into the egg cell itself.

Endometrium
Greek
endo- inside, within
-metra- womb
-y place for an activity; condition, state
Mucous membrane lining the interior surface of the uterus.

Endomorphic
Greek
endo- inside, within
-morph- shape, form, figure, or appearance
-ic (ikos) relating to or having some characteristic of
An individual characterized by a significant amount of soft tissue around the area of the abdomen; this fatty tissue develops from the embryonic endodermal layer.

Endoplasm
Greek
endo- inside, within
-plasm (plassein) to mold or form cells or tissues
A central, less viscous portion of the cytoplasm that is distinguishable in certain cells, especially motile cells.

Endopod
Greek
endo- inside, within
-podos foot
Medial branch of a biramous crustacean appendage.

Endorphin
Greek
endo- inside, within
-morpheus- god of dreams
-in protein or derived from a protein.
A morphine-like substance secreted in the pituitary gland to control pain and pleasure.

Endoskeleton
Greek
endo- inside, within
-skeletos hard

A supporting framework within the living tissues of an organism.

Endosperm
Greek
endo- inside, within
-sperma seed
In flowering plants, storage tissue.

Endospore
Greek
endo- inside, within
-spora seed
A small asexual spore that develops inside the cell of some bacteria and algae.

Endostyle
Greek
endo- inside, within
-sylos a pillar
Ciliated groove(s) in the floor of the pharynx of tunicates, cephalochordates, and larval cyclostomes, used for accumulating and moving food particles to the stomach.

Endothermal
Latin/Greek
endo- inside, within
-thermos- combining form of "hot" (heat)
-al of the kind of, pertaining to, having the form or character of
Pertaining to chemical reactions in warm-blooded animals that generate heat for the maintenance of a constant internal environment.

Endothermic
Greek
endo- inside, within
-thermos- combining form of "hot" (heat)
-ic (ikos) relating to or having some characteristic of
Characterized by or causing the absorption of heat.

Energy
Greek
en- in, at, onto
-ergon work
The capacity to do work; source of usable power; vigorous exertion of effort.

Enneagynous
Greek
ennea- nine
-gynous in relation to the female organ of a plant
In botany, having nine pistils or styles in a flower.

Enterocoel
Greek
enteron- gut
-koiloma cavity

A type of coelom formed by the outpouching of a mesodermal sac from the endoderm of the primitive gut.

Enterocoelomate
Greek
enteron- gut
-koiloma- cavity
-ate of or having to do with
An animal having an enterocoel, such as an echinoderm or a vertebrate.

Enthalpy
Greek
en- in, at, onto
-thalpien- to heat
-y place for an activity; condition, state
The sum of the internal energy of a body and the product of its volume multiplied by its pressure.

Entomology
Greek
entomos- cut from two, segmented
-logy (logos) used in the names of sciences or bodies of knowledge
The scientific study of insects.

Entropy
Greek
en- in, at, onto
-trope transformation
The tendency for all matter and energy in the universe to evolve toward a state of inert uniformity.

Environmentalist
French
environ- round about; encircle
-ment- state or condition resulting from a (specified) action
-al- of the kind of, pertaining to, having the form or character of
-ist agent, specialist
A person who seeks to protect the natural environment.

Enzyme
Greek
en- in, at, onto
-zume ferment, leaven
Produced by living cells that catalyze chemical reactions in organic matter.

Eocene
Greek
eos- dawn
-kainos recent
An epoch of the lower Tertiary period, spanning the time between 55.5 and 33.7 million years ago.

Eon
Greek
aion indefinitely long period of time
Longest period of geologic time.

Eosinophil
Greek
eos- dawn (color of), rose, red
-in- protein or derived from a protein
-phile one who loves or has a strong affinity or preference for
A granular bilobed leukocyte with coarse cytoplasmic granules that attract the red acid dye eosin, a biological stain for studying cell structures.

Ephemeroptera
Greek
ephemeros- for a day
-pteron wing
Mayflies; fragile winged insects that develop from aquatic nymphs and live as adults for only a few days.

Epibenthos
Greek
epi- above, over, on, upon
-benthos deep; the fauna and flora of the bottom of the sea
The community of organisms living at the surface of the seabed or lake floor.

Epiblast
Greek
epi- above, over, on, upon
-blastos bud, germ cell
The outer layer of the blastula giving rise to the ectoderm.

Epicardium
Greek
epi- above, over, on, upon
-kard- heart, pertaining to the heart
-ium quality of the relationship
The inner layer of the pericardium, a conical sac of fibrous tissue that surrounds the heart.

Epicenter
Greek
epi- above, over, on, upon
-kentron center, sharp point
The point of the earth's surface directly above the focus of an earthquake.

Epicycle
Greek
epi- above, over, on, upon
-kyklos circle, wheel, cycle
A circle whose circumference rolls along the circumference of a fixed circle.

Epidemic

Greek

epi- upon, above

-demos- people

-ic (ikos) relating to or having some characteristic of

A disease found among many people in an area; a situation where an infectious disease develops and spreads quickly through a population.

Epidendrous

Greek

epi- above, over, on, upon

-dendr- tree, treelike structure

-ous full of, having the quality of, relating to

Relating to organisms that grow or exist on trees.

Epidermis

Greek

epi- above, over, on, upon

-dermis skin

The outer epithelial layer of the external integument of the animal body that is derived from embryonic epiblast.

Epididymis

Greek

epi- above, over, on, upon

-didumos twins, testicles

Long, narrow, convoluted tube on the top, posterior aspect of either of the two testes; it is part of the sperm duct system.

Epigastrium

Greek

epi- above, over, on, upon

-gastr- stomach, belly

-ium quality of the relationship

The part of the abdominal wall lying on or over the stomach.

Epiglottis

Greek

epi- above, over, on, upon

-glotta tongue

The thin elastic cartilaginous structure located at the root of the tongue that folds over the glottis to prevent food and liquid from entering the trachea during the act of swallowing.

Epinephrine

Greek

epi- above, over, on, upon

-nephros- kidneys

-ine a chemical substance

An endogenous adrenal hormone that increases cardiac activity, dilates bronchial tubes, and stimulates the production of glucose from glycogen.

Epiphyseal (line)

Greek

epi- above, over, on, upon

-phyein- to grow

-al of the kind of, pertaining to, having the form or character of

Pertaining to or resembling the epiphysis; in long bone development; the line that results when the ossification process of the shaft meets with the bony development at the end of a bone.

Epiphyte

Greek

epi- above, over, on, upon

-phuton plant having a (specified) characteristic or habitat

A plant, such as a tropical orchid or a staghorn fern, that grows on another plant upon which it depends for mechanical support but not for nutrients; also called aerophyte, air plant.

Epipod

Greek

epi- above, over, on, upon

-pous podos, foot

A lateral process on the protopod of a crustacean appendage often modified as a gill.

Episode

Greek

epi- above, over, on, upon

-eisodios coming in besides, entering

An incident or event that stands out from the continuity of everyday life.

Episome

Greek

epi- above, over, on, upon

-soma (somatiko) body

A genetic unit or gene that has the capacity to exist outside of or independently of its host cell.

Epistasis

Greek

epi- above, over, on, upon

-histanai- to place; to stop

-sis action, process, state, condition

The suppression of a bodily discharge such as urine. In genetics, the suppression of the expression of a gene by another gene.

Epistome

Greek

epi- above, over, on, upon

-stoma mouth

Flap over the mouth in some lophophorates that bears the protocoel.

Epithethia
Greek
epi- above, over, on, upon
-thele- nipple
-ia names of diseases, place names, or Latinizing plurals
Papillary projections of the epithelium that penetrate the underlying stroma of connecting tissue.

Epitope
Greek
epi- above, over, on, upon
-topos place, spot
A portion of a protein molecule that is the specific target of an immune response.

Epizootic
Greek
epi- above, over, on, upon
-zoon- animal, animal-like
-ic (ikos) relating to or having some characteristic of
Affecting a large number of animals at the same time within a particular region or geographic area; used in reference to a disease.

Epoch
Greek
ep- time
-och fixed
Subdivision of a period on the geologic time scale.

Equation
Latin
aequi- equal, same, similar, even
-ion state, quality, or process of
A representation of a chemical reaction, usually written as a linear array in which the symbols and quantities of the reactants are separated from those of the products by an equal sign, an arrow, or a set of opposing arrows.

Equator
Latin
aequi- equal, same, similar, even
-or from
The imaginary great circle around the earth's surface, equidistant from the poles and perpendicular to the earth's axis of rotation; it divides the earth into the Northern Hemisphere and the Southern Hemisphere.

Equilibrate
Latin
aequi- equal, same, similar, even
-libr- balanced, level; make even; weight
-ate characterized by having
Having to maintain in or bring into equilibrium.

Equilibrium
Latin
aequi- equal, same, similar, even
-libr- balanced, level; make even; weight
-ium quality or relationship
A state of balance between opposing forces or actions.

Equine
Latin
equus- horse
-ine of or relating to
Of or belonging to the family Equidae, which includes the horses, asses, and zebras.

Equinox
Latin
aequi- equal, same, similar, even
-noct night
Either of the two times during a year when the sun crosses the celestial equator and when the day and night are approximately equal in length.

Equipollent
Latin
aequi- equal, same, similar, even
-pollere- to be powerful
-ent causing an action; being in a specific state
Equal in force, power, effectiveness, or significance.

Equipotential
Latin
aequi- equal, same, similar, even
-potent- power; to be able
-ial (variation of *-ia*) relating to or characterized by
The work required to move a unit of positive charge, a magnetic pole, or an amount of mass from a reference point to a designated point in a static electric, magnetic, or gravitational field; potential energy.

Era
Latin
aera counters
The longest of the geological time periods, usually marked by some catastrophic geological event.

Eremic
Greek
erem- lonely, solitary; hermit; desert
-ic (ikos) relating to or having some characteristic of
Pertaining to deserts or sandy regions.

Eremobiology
Greek
erem- lonely, solitary; hermit; desert
-bios- life, living organisms or tissue
-logy (logos) used in the names of sciences or bodies of knowledge
The science of biology in arid ecological systems.

Eremophile
Greek
erem- lonely, solitary; hermit; desert
-phile one who loves or has a strong affinity or preference for
Organisms that survive and thrive in desert or desertlike conditions.

Eremophyte
Greek
erem- lonely, solitary; hermit; desert
-phuton plant having a (specified) characteristic or habitat
A plant species that has developed the adaptations to live in arid, desertlike conditions.

Erg
Greek
ergon work
A small unit of work equal to the force of one dyne acting over a distance of one centimeter.

Ergonomics
Greek
ergon- work
-nom- (nemein) to dictate the laws of; knowledge; usage; order
-ic (ikos) relating to or having some characteristic of
The applied science of equipment design, as for the workplace, intended to maximize productivity by reducing operator fatigue and discomfort.

Erogenous
Latin
eros- sexual love or sexual passion
-gen- to give birth, kind, produce
-ous full of, having the quality of, relating to
Producing erotic feelings; often a reference to parts of the body that are sensitive to sexual arousal.

Erosion
Latin
erosio- an eating away
-ion state, process, or quality of
The group of natural processes, including weathering, dissolution, abrasion, corrosion, and transportation, by which material is worn away from the earth's surface.

Eruciform
Latin
eruci- caterpillar
-forma having the form of
Applied to insect larvae, caterpillar-like; more or less cylindrical with a well-developed head and stumpy legs at the rear, in addition to the true thoracic legs. The caterpillars of butterflies and moths are typical examples.

Erythroblast
Greek
eruthros- red
-blastos bud, germ cell
Immature red blood cells found within the red bone marrow of mammals; they are typically nucleated.

Erythroblastosis
Greek
eruthros- red
-blastos bud, germ cell
-osis increase, formation
An abnormal presence of immature red blood cells in the bloodstream.

Erythrocyte
Greek
eruthros- red
-cyte (kutos) sac or bladder that contains fluid
Red blood cell that contains hemoglobin and carries oxygen from the lungs or gills to the tissues in vertebrates.

Erythropoiesis
Greek
eruthros- red
-poiein- production, formation; to make
-sis action, process, state, condition
The process of the production of red blood cells in the red bone marrow.

Erythropoietin
Greek
eruthros- red
-poiein- production, formation; to make
-in protein or derived from protein
A chemical secreted by the kidney to regulate the production of red blood cells.

Esophagoduodenostomy
Greek/Latin
ois- (pherein) to carry
-phagos- (phagein) to eat, eating
-duodeni- twelve each
-stoma- opening
-y place for an activity; condition, state
Surgical removal of the stomach, followed by connection of the esophagus to the duodenum.

Esophagus
Greek
ois- (future tense of *pherein*) to carry
-phagos- (phagein) to eat; eating
-us thing
A muscular, membranous tube extending from the pharynx to the stomach.

Ester
German (from Latin)
essig vinegar
Any of a class of organic compounds corresponding to the inorganic salts and formed from an organic acid and an alcohol.

Esterification
Greek
äther- etherlike acid
-fication action, process, or quality of
A reaction involving a group of organic compounds that causes the reagents (usually a carboxylic acid and alcohol) to become an ester.

Estivation
Latin
estiv- dormancy in the summer
-ion state, process, or quality of
The process of spending the summer in a resting state.

Estrogen
Greek
oistros- frenzy; gadfly
-gen to give birth, kind, produce
Female sex hormones secreted by both the ovaries and the adrenal cortex.

Estuary
Latin
aestus- tide, surge
-ary of, relating to, or connected with
An arm of the sea that extends to meet the mouth of a river.

Ethane
Greek
eth- organic functional group with two carbons
-ane organic compound containing no multiple bonds
An odorless alkane gas, C_2H_6.

Ether
Greek
aither upper air
Any of a class of organic compounds in which two hydrocarbon groups are linked by an oxygen atom.

Ethnobotany
Greek
ethnos- people or races
-botanē- fodder, plants
-onuma name
The study of the relationship between humans and plants.

Etiology
Greek
aitia- cause
-logy (logos) used in the names of sciences or bodies of knowledge
The scientific study of the causes and origins of diseases.

Etymology
Greek/Latin
etymon- true sense; earlier form of a word
-logy (logos) used in the names of sciences or bodies of knowledge
The study of the sources and development of words.

Eubacteria
Greek
eu- good, well; true
-bacter- microscopic organism
-baktron- staff, rod
-ia names of diseases, place names, or Latinizing plurals
Large group of bacteria having rigid cell walls.

Euglena
Greek
eu- good, well; true
-glene eyeball
Any organism of the genus *Euglena*, found in freshwater and characterized by chlorophyll, a single flagellum, and a reddish "eyespot."

Euhaline
Greek
eu- good, well; true
-hal- salt
-ine in a chemical substance
Term used with reference to normal sea water, containing 30 to 40 parts per thousand salt; applies to organisms thriving in this environment.

Eukaryote
Greek
eu- good, well; true
-kairon nut; cell nucleus
An organism whose cells contain a distinct, membrane-bound "true" nucleus.

Eumetazoans
Latin
eu- good, well; true
-meta- later in time
-zoan animal
Animals with both tissues and symmetry.

Euphotic (zone)
Greek
eu- good, well; true
-photos- light, radiant energy
-ic (ikos) relating to or having some characteristic of
Of, relating to, or being the uppermost layer of a body of water that receives sufficient light for photosynthesis and the growth of green plants.

Eupnea

New Latin
eu- normal
-pnion breathing or breath
Normal, rhythmic, unlabored breathing rates.

Eurybaric

Greek
eury- wide, broad
-bar- weight, pressure
-ic (ikos) relating to or having some characteristic of
Applicable to animals adaptable to great differences in altitude.

Euryhalic

Greek
eury- wide, broad
-hal- salt
-ic (ikos) relating to or having some characteristic of
Able to tolerate a wide range of salinity; said of organisms capable of withstanding widely varying concentrations of salt in the environment.

Euryhaline

Greek
eury- wide
-hal- salt
-ine in a chemical substance
Able to tolerate wide ranges of saltwater concentrations.

Euryphagous

Greek
eury- wide
-phagos- (phagein) to eat, eating
-ous full of, having the quality of, relating to
An ecological term referring to an organism that eats a large variety of foods.

Euryphotic

Greek
eury- wide, broad
-phot- light
-ic (ikos) relating to or having some characteristic of
Tolerant of a wide range of light intensity, typically measured between a forest and a field.

Eurypterid

Greek
eury- wide
-pteron- wing
-id state, condition; having, being, pertaining to, tending to, inclined to
Large, extinct scorpion-like arthropod considered to be related to horseshoe crabs.

Eurytopic

Greek
eury- wide
-topos place
-ic (ikos) relating to or having some characteristic of
Refers to an organism or species capable of living within a wide environmental range.

Eutrophic

Greek
eu- good, well, true
-trophos- (trophein) to nourish; food, nutrition; development
-ic (ikos) relating to or having some characteristic of
Having waters rich in mineral and organic nutrients, causing plant life to proliferate, thereby reducing the dissolved oxygen content and often killing off other organisms.

Eutrophication

Greek
eu- good, well; true
-trophos- (trophein) to nourish; food, nutrition; development
-ation action, process, or quality of
The process by which a body of water becomes enriched in dissolved nutrients (such as phosphates) that stimulate the growth of aquatic plant life, usually resulting in the depletion of dissolved oxygen.

Evacuate

Latin
-vacare- empty
-ate of or having to do with
To empty or send away; to eliminate or excrete wastes from a living body.

Evagination

Latin
-vagina- sheath
-ion state, process, or quality of
An outpocketing from a hollow structure; to turn a body part inside out.

Evaporation

Latin
vaporatus- steam, vapor
-ion state, process, or quality of
Vaporization of a liquid below its boiling point.

Evapotranspiration

Latin
ex- outside, outward, out of, out; away from
-vaporatus- steam, vapor
-trans- across or through
-spirāre- to breath
-ion state, process, or quality of
The sum total of water loss due to evaporation and plant transpiration.

Evolution
Latin
evolut- unrolling
-ion state, process, or quality of
The theory that the various types of animals and plants have their origin in other, preexisting types and that the distinguishable differences are due to modifications in successive generations.

Excision
Greek
ex- outside, outward, out of, out; away from
-cis- to cut
-ion state, process of
The process of cutting off something small by surgery.

Excited
Latin
ex- outside, outward, out of, out; away from
-ciere to set in motion
Being at an energy level higher than the ground state.

Excretion
Latin
ex- outside, outward, out of, out; away from
-cernere- to separate
-ion state, process of
To separate and eliminate or discharge (waste) from the blood or tissues or from active protoplasm.

Exfoliate
Latin
ex- outside, outward, out of, out; away from
-folium- leaf
-ate of or having to deal with
To come off or separate into flakes, scales, or layers; mechanical weathering process in which outer rock layers are stripped away, often resulting in dome-shaped formations.

Exobiology
Greek
ex- outside, outward, out of, out; away from
-bios- life, living organisms or tissue
-logy (logos) used in the names of sciences or bodies of knowledge
Study of life forms that possibly exist elsewhere in the universe.

Exocytosis
Greek
ex- outside, outward, out of, out; away from
-cyte- (kutos) sac or bladder that contains fluid
-sis action, process, state, condition
The process of moving things to the outside of a cell.

Exopod
Greek
ex- outside, outward, out of, out; away from
-podos foot
Lateral branch of a biramous crustacean appendage.

Exoskeleton
Greek
ex- outside, outward, out of, out; away from
-skeletos dried up (body)
A hard outer structure, such as the shell of an insect or crustacean, that provides protection or support for an organism.

Exosphere
Greek
ex- outside, outward, out of, out; away from
-sphaira a globe shape, ball, sphere
The outer layer of the thermosphere, extending into space.

Exothermal
Greek
ex- outside, outward, out of, out; away from
-thermos- combining form of "hot" (heat)
-al of the kind of, pertaining to, having the form or character of
Characterized by or formed with the evolution of heat.

Exothermic
Greek
ex- outside, outward, out of, out; away from
-thermos- combining form of "hot" (heat)
-ic (ikos) relating to or having some characteristic of
Referring to a chemical reaction where heat is released from the source.

Exotic
Greek
ex- outside, outward, out of, out; away from
-otic state or condition of; condition of being
Strikingly, excitingly, or mysteriously different or unusual; from another part of the world.

Expedition
Greek
ex- outside, outward, out of, out; away from
-pedi- foot
-ion state, process, or quality of
A journey or excursion undertaken for a specific purpose.

Experiment
Latin
experiri- to try
-ent causing an action or being in a specific state
A test under controlled conditions that is made to demonstrate a known truth, examine the validity of a hypothesis, or determine the efficacy of something previously untried.

Holland in the Seventeenth Century

Able to form a republic in the seventeenth century by declaring its independence from Spain, Holland was left to its own resources to either flourish or decline. Thus, the economy of Holland was dependent on the free-thinking, creative society of its day. Beginning in that century, but associated more with the eighteenth century, was the Age of Enlightenment, a period characterized by reason rather than the traditions of the Dark Ages. This movement led to an unparalleled optimism and to bold expressions of philosophy, law, art, science, and government. The Dutch embraced the Age of Enlightenment, which eventually spread throughout Europe.

The formation of the Dutch East India Company required the recruitment of skilled craftsmen to build a fleet of ships capable of traveling great distances. The Dutch sailor-merchants sailed all over the world and brought back the rarest of goods for sale. Exploration became a part of the social fiber of the Dutch people. Science, mathematics, and philosophy flourished in Holland, where all free thinkers were welcome to explore their passions. There was little to fear from the Church, which still held a grip over much of Europe. Men feared for their lives when scientific reason clashed with the accepted Church dogma. Thus seventeenth-century Holland became home to many migrating scientists and others who sought freedom to express their ideas. In Amsterdam Anton Van Leeuwenhoek, known as the father of microbiology, invented the microscope during this period. It is said that his microscopes, equipped with lenses that he himself ground, were able to magnify well over 500 times normal vision. Only a handful of the hundreds of microscopes he crafted still exist today.

Christian Huygens crafted lenses for telescopes and created a telescope that was over 5 meters long. He speculated that the atmosphere of Venus caused the planet to be covered by clouds. He observed the patterns of rotation of planets, and he estimated quite accurately the length of a Martian day. Huygens was the first to recognize the rings of Saturn, and he also discovered Titan, the planet's largest moon. These are only a few of the incredible discoveries and inventions this scientist is responsible for.

Countless people have been inspired over the ages by this colony's many explorers, adventurers, craftsmen, statesmen, artists, mathematicians, and philosophers. Even Albert Einstein was influenced by a Portuguese-Jewish philosopher who lived in Holland, Benedict (Baruch) Spinoza.

Exsiccated
Latin
ex- outside, outward, out of, out; away from
-sicca- drying
-ate characterized by having
Dried, especially in reference to soils that have lost their moisture.

Extensor
Greek
ex- outside, outward, out of, out; away from
-ten- to move in a certain direction; to stretch, hold out
-or a condition or property of things or persons; person who does something
Any of various muscles that extend or straighten some part of the body, especially a flexed arm or leg.

External
Latin
externus- outward
-al pertaining to, having the form or character of
Relating to, existing on, or connected with the outside or an outer part; exterior.

Extinction
Latin
ex- outside, outward, out of, out; away from
-stinguere- to quench
-ion state, process, or quality of
Ceasing of existence of a species.

Extraction
Greek
ex- outside, outward, out of, out; away from
-trahere- to draw
-ion state, process, or quality of
To obtain from a substance by chemical or mechanical action, as by pressure, distillation, or evaporation.

Extrusive
Latin
ex- outside, outward, out of, out; away from
-trudere thrust
Igneous rock that forms when molten rock solidifies above the surface.

Eye
Modern English
eghe resembling an eye shape
The development of a calm center of a storm.

F

Famine
> Latin
> *fames-* hunger
> *-ine* of or relating to
> A drastic, wide-reaching food shortage threatening the lives of an entire population.

Fault
> Latin
> *fallere* to deceive, fail
> To shift so as to produce a fault.

Fecundity
> Latin
> *fecund-* fruitful, fertile
> *-ity* state of, quality of
> Refers to female animals: the faculty of reproduction; the capacity for bringing forth young; productiveness. In botany, the faculty or power of germinating.

Fermentation
> Latin
> *fermentum-* splits complex organic compounds into simpler ones
> *-ion* state, process or quality of
> A type of anaerobic pathway of ATP formation: it starts with glycolysis, ends when electrons are transferred back to one of the breakdown products or intermediates, and regenerates the NAD+ required for the reaction. Its net yield is two ATP per glucose molecule degraded.

Ferroalloy
> Latin
> *ferrum-* iron; pertaining to, or containing iron
> *-alligare* to bind

Any of various alloys of iron and one or more other elements.

Ferrotherapy
> Latin
> *ferrum-* iron; pertaining to, or containing iron
> *-therapeuein* to heal, cure; treatment
> The treatment of disease with iron.

Fertilization
> Latin
> *fertilis-* to bear
> *-ion* state, process, or quality of
> The act or process of initiating the reproductive process in sexual creatures by the union of an egg and a sperm cell.

Fibrin
> Latin
> *fibro-, fibr-, fibra-* fiber; an elongated threadlike structure
> *-in* protein or derived from protein
> Large insoluble strands of protein that aid in the clotting of blood.

Fibrinogen
> Latin/Greek
> *fibro-, fibr-, fibra-* fiber; an elongated threadlike structure
> *-gen* to give birth, kind, produce
> A blood plasma protein that turns into fibrin when converted by thrombin during the blood-clotting process.

Fibronectin
> Latin/ Greek
> *fibro-, fibr-, fibra-* fiber; an elongated threadlike structure

-nhkto- **(Greek)** swimming
-in protein or derived from protein
A fibrous linking protein that functions as a reticuloendothelial mediated host defense mechanism and is impaired by surgery, burns, infection, neoplasia, and disorders of the immune system.

Fibrosis
Latin
fibro-, fibr-, fibra- fiber; an elongated threadlike structure
-sis action, process, state, condition
The formation of excess fibrous tissue, usually as an attempt to repair damaged tissue or as a reaction to a trauma.

Field
Old English
feld field
A region of space characterized by a physical property, such as gravitational or electromagnetic force or fluid pressure, having a determinable value at every point in the region.

Filial
Latin
fili- son, daughter, offspring
-al of the kind of, pertaining to, having the form or character of
Of or relating to a generation or the sequence of generations following the parental generation.

Filipodium
Latin
filum- thread
-podos- foot
-ium quality or relationship
A type of pseudopodium that is very slender and may branch, but does not rejoin to form a mesh.

Filtration
Latin
filtrum- to put or go through a filter
-ion state, process, or quality of
A process in which mixtures are separated based upon the size of particles that can fit through a filter.

Fimbriae
Latin
fimbriae thread, fringe
A thread or fringelike anatomical part of an organ, such as the aperture to the Fallopian tubes.

Fine
Latin
finis utmost limit, end

In chemistry, refers to having a stated amount of gold or silver in it. A gold or silver alloy that is 925/1000 fine is 92.5% gold or silver.

Fission
Latin
fissus- splitting
-ion state, process, or quality of
Act or process of splitting or breaking up into parts.

Fistula
Latin
fistula pipe
An abnormal duct or canal resulting from injury, disease, or congenital disorder that extends from the hollow of a body organ to the surface or to another organ.

Fixation
Latin
fixus- to fasten
-ation action, process, or quality of
The process of conversion into a more reactive, usable form.

Fjord
Old Norse
fjordhr inlet
A long, narrow, deep inlet of the sea between steep slopes.

Flagellum
Latin
flagrum whip
A long, threadlike appendage; a whiplike extension.

Flammable
Greek
philogiston flammable
Describes a substance that is easily ignited and capable of burning.

Flexor
Latin
flectere- to bend
-or a condition or property of things or persons; person who does something
Any muscle that bends a limb.

Flocculate
Latin
flocculus- tuft
-ate of or having to do with
To form into woolly, soft, or cloudlike masses; to form compound masses, as a cloud or a chemical precipitate.

Flood
Middle English
flud flowing water

The overflowing of water on land that is usually dry; a deluge.

Fluctuate
Latin
fluere- to flow, wave
-ate of or having to do with
To vary irregularly; to rise and fall in waves.

Fluid
Latin/Greek
fluere- to flow, wave
-id state, condition; having, being, pertaining to, tending to, inclined to
A continuous, amorphous substance whose molecules move freely past one another and that has the tendency to assume the shape of its container; a liquid or gas.

Fluke
Greek
plax flat surface
A flattened, digenetic trematode worm.

Fluorescence
Latin
fluere- to flow, wave
-escentia state or process of
The process in which an atom releases energy in the form of electromagnetic radiation.

Fluoroscope
Latin/Greek
fluere- to flow, wave
-skopion for viewing with the eye
An imaging device using x-rays to project a fluorescent image on a screen.

Fluvial
Latin
fluvi- river, stream
-al of the kind of, pertaining to, having the form or character of
Pertaining to rivers and river activities; found or living in a river; produced by a river or stream.

Fluvioterrestrial
Latin
fluvi- river, stream
-terra- of or relating to the earth or its inhabitants
-ial of or relating to
Refers to inhabiting streams and the surrounding land.

Flux
Latin
fluxus (past participle of *fluere*) to flow
The rate of flow of fluid, particles, or energy through a given surface.

Foliaceous
Latin
folium- leaf
-aceous of or relating to a plant family
Belonging to, or having the texture or nature of foliage or leaves; leaflike in form or made of growth; composed of thin laminated layers, as certain rocks.

Foraminiferan
Latin
forare- to bore; hole, an opening,
-ferre to bear
A member of the class Granuloreticulosea bearing a shell with many openings.

Forbicolous
Greek
pherbein- to graze
-cola tiller, inhabitant
Living on broad-leaved plants; herbicolous.

Forbivorous
Greek/Latin
pherbein- to graze
-vorare- swallow, devour
-ous full of, having the quality of, relating to
Feeding on broad-leaved plants.

Force
Latin
fortis strong
A vector quantity that tends to produce an acceleration of a body in the direction of its application.

Forensic
Latin
forensis- public
-ic (ikos) relating to or having some characteristic of
Relating to or dealing with the application of scientific knowledge to legal problems.

Forest
Latin
foris outside
A dense growth of trees, plants, and underbrush covering a large area.

Formation
Latin
format- shape, figure, appearance
-ion state, process, or quality of
The act or process of arranging something or of taking form.

Formicary
Latin
formic- ant
-ary of, relating to, or connected with
A nest of ants or anthill.

Fossil
Latin
fossilis dug up
Having the characteristics of a fossil: preserved in a mineralized or petrified form from a past geologic age.

Fractal
Latin
frangere- to break
-al of the kind of, pertaining to, having the form or character of
A geometric pattern that is repeated at ever smaller scales to produce irregular shapes and surfaces that cannot be represented by classical geometry.

Fractionate
Latin
frangere- to break
-ate of or having to do with
To separate a mixture by distillation, crystallization, or other method into its ingredients or into portions that have different properties.

Fractoluminescence
Latin
frangere- to break
-lumen- light
-ence the condition of
The emission of light from the fracture of a crystal.

Frequency
Latin
frequens- a crowd, throng
-cy state, condition, quality
The number of wave peaks occurring in a unit of time.

Friction
Latin
fricare- to rub
-ion state, process, or quality of
The force generated opposite to the motion of an object resulting from an interaction of surfaces.

Frigid
Latin
frigus- cold, frost
-id state, condition; having, being, pertaining to, tending to, inclined to
Refers to extreme cold, with a very cold temperature.

Fructose
Latin
fructus- fruit
-ose sugar, carbohydrate
A very sweet sugar occurring in many fruits and honey and used as a preservative for foodstuffs and as a intravenous nutrient.

Fruit
Latin
fructus fruit
The ripened ovary or ovaries, together with accessory parts, containing the seeds of a seed-bearing plant and occurring in a wide variety of forms.

Fucivorous
Greek/Latin
phukos- rock lichen, seaweed
-vorare- to swallow, devour
-ous full of, having the quality of, relating to
Feeding or subsisting on seaweed and related sea and ocean foods.

Fulcrum
Latin
fulcire to support
The point or support on which a level pivots.

Fumaroles
Latin
fumus- smoke, vapor
-ole little
A crack or fissure that releases gases from a volcano.

Fumatorium
Latin
fumus- smoke, vapor
-ate- to do, to make, to cause
-orium a place or a thing used for something
An airtight compartment in which vapor may be generated to destroy germs or insects.

Fume
Latin
fumus smoke, vapor
Vapor, gas, or smoke, especially if harmful, strong, or odorous.

Function
Latin
fungi- to do, perform, execute, discharge
-ion state, process, or quality of
The special, normal, or proper physiological activity performed by an organ or part.

Fundamental
Latin
fundus- bottom
-ment- state or condition resulting from a (specified) action
-al of the kind of, pertaining to, having the form or character of
Of or relating to the foundation or base.

Fungal

Latin

spongos- spongelike

-al of the kind of, pertaining to, having the form or character of

Caused by a fungus, or relating to or having the characteristics of a fungus.

Fungicide

Greek/Latin

spongos- spongelike

-cide (caedere) to cut, kill, hack at, or strike

The destruction of fungi or something used to kill fungi (spores).

Fungus

Greek

spongos- spongelike

-us singular

Eukaryotic organisms lacking chlorophyll and vascular tissue. They range from unicellular to multicellular. Many produce fruiting bodies.

Fusion

Latin

fundere- to melt

-ion state, process, or quality of

The joining into a single entity.

G

Galactose
Greek
galakt- milk
-ose sugar, carbohydrate
$C_6H_{12}O_6$; one of the hextose sugars, it is found in pectins and gums.

Galaxy
Greek
galakt- milk
-ia names of diseases, place names, or Latinizing plurals
Any of numerous large-scale aggregates of stars, gas, and dust that constitute the universe, containing an average of 100 billion (10^{11}) solar masses and ranging in diameter from 1,500 to 300,000 light-years. Also called nebula.

Gallbladder
Old English
galla- nutgall
-blaēdre bladder
A small, hollow, saclike, muscular organ located below the liver. It contains bile that is produced by the liver and secretes the bile into the small intestine to aid in the digestion of fats.

Gallimimus
Latin
gallus- rooster
-mimus mimic
A dinosaur whose fossil remains resemble a very large rooster and that existed during the Late Cretaceous period in Mongolia.

Gametangium
Greek/Latin
gamet- husband or wife; to marry
-angeion- vessel
-ium quality or relationship
The reproductive organ of bryophytes, consisting of the male antheridium and the female archegonium; a multichambered jacket of sterile cells in which gametes are formed.

Gamete
Greek
gamein to marry
Either a male or female reproductive cell possessing the haploid number of chromosomes.

Gametocyte
Greek
gamet- husband or wife; to marry
-cyte (kutos) sac or bladder that contains fluid
The mother cell of a gamete; that is, an immature gamete.

Gametogenesis
Greek
gamet- husband or wife; to marry
-gen- to give birth, kind, produce
-sis action, process, state, condition
The process in which production of gametes, eggs or sperm, occurs.

Gametophyte
Greek
gamet- husband or wife; to marry
-phyte a plant
A stage in a plant's life cycle during which eggs and sperm are produced.

Ganglia
Greek
gangl- nerve bundle
-ia names of diseases, place names, or Latinizing plurals
Masses of nerve tissue containing nerve cells external to the brain or spinal cord.

Gangue
French (from German)
gang lode
Worthless rock or other material in which valuable minerals are found.

Gas
Greek
chaos empty, space
Matter that has no fixed volume or shape; it conforms to the volume and shape of its container.

Gastrectomy
Greek
gastr- stomach, belly
-ekt- outside, external, beyond
-tomos (temnein) to cut, incise, section
Cutting out or removing the stomach.

Gastric
Greek
gastr- stomach
-ic (ikos) relating to or having some characteristic of
Pertaining to or having some characteristic of the stomach.

Gastrodermis
Greek
gastr- stomach, belly
-derma skin
Lining of the digestive cavity of cnidarians.

Gastroenteritis
Greek
gastr- stomach, belly
-enteron- small intestine
-itis inflammation, burning sensation
Inflammation of the mucous membrane of the stomach and intestines.

Gastromalacia
Greek
gastr- stomach, belly
-malacia softening of tissue
Softening of the walls of the stomach, usually occurring after death.

Gastromegaly
Greek
gastr- stomach, belly
-megaly large
Enlargement of the abdomen or the stomach.

Gastroplexy
Greek
gastr- stomach, belly
-plexy fixation
Fixation of the stomach.

Gastropod
Greek
gastr- stomach, belly
-podos foot
Any of a group of mollusks that have a broad disk-like organ of locomotion on the ventral surface of the body.

Gastroptosis
Greek/Latin
gastr- stomach, belly
-ptosis downward, displacement, drooping, saggy
Downward displacement of the stomach.

Gastrovascular
Greek/Latin
gastr- stomach, belly
-vas- vessel, duct
-cul- small, tiny
-ar relating to or resembling
Describes the primary organ of coelenterates that functions both in digestion and in the transportation of nutrients to all parts of an animal's body.

Gastrula
Greek
gastr- stomach, belly
-ula diminutive
An embryo at the stage following the blastula, consisting of a hollow, two-layered sac of ectoderm and endoderm surrounding an archenteron that communicates with the exterior through the blastopore.

Gemmules
Latin
gemma- bud
-ule little, small
Asexual, cystlike reproductive unit in freshwater sponges; formed in summer or autumn and capable of overwintering.

Genetic
Greek
gen- origin, birth
-ic (ikos) relating to or having some characteristic of
The branch of biology that deals with heredity, especially the mechanisms of hereditary transmissions and the variation of inherited characteristics among similar or related organisms; the genetic makeup of an individual, a group, or a class.

Genome
Greek
gen- origin, birth
-ome group
Total number of genes in an individual.

Genotype
Greek
gen- origin, birth
-typos mark
The complete genetic constitution of an organism or group as determined by the specific combination and location of the genes on the chromosome.

Genus
Latin
genus race
A group of related species with taxonomic rank between family and species.

Geobios
Greek
ge- earth, world
-bios life, living organisms, or tissue
The total life of the land; that part of the earth's surface occupied by terrestrial organisms; terrestrial life.

Geocentric
Greek
ge- earth, world
-kentron- a point or place that is equally distant from the sides or outer boundaries of something; the middle
-ic (ikos) relating to or having some characteristic of
Refers to early accepted position by scientists/philosophers that the earth was the center of the solar system and that all objects in the sky revolved around the earth.

Geodesic
Greek
ge- earth, world
-daiesthai to divide
Describes the path an object will follow through space and time in the absence of external forces.

Geography
Greek
ge- earth, world
-graphia (graphein) to write, record, draw, describe
The study of the earth and its features and of the distribution of life on the earth, including human life and the effects of human activity.

Geology
Greek
ge- earth, world
-logy (logos) used in the names of sciences or bodies of knowledge
Of or relating to the study of the earth, including soils, mineralogy, and the dynamics of the earth's crust.

Geonyctitropism
Greek
ge- earth, world
-nycto- night; a relationship to darkness, dark
-trope- bend, curve, turn, a turning; response to stimulus
-ium quality or relationship
Orientation movements in plants during darkness in response to gravity.

Geophysiology
Greek
ge- earth, world
-phusio- form, origin, nature
-logy (logos) used in the names of sciences or bodies of knowledge
The study of the interaction among all organisms living on the earth.

Geosynchronous
Greek
ge- earth, world
-synchron- at the same time
-ous full of, having the quality of, relating to
Refers to a geocentric orbit that has the same orbital period as the sidereal rotation period of the earth.

Geothermal
Greek
ge- earth, world
-therm- heat, hot, warm
-al of the kind of, pertaining to, having the form or character of
Of, relating to, or using the heat of the earth's interior; also, to be produced or permeated by such heat.

Germination
Latin
germinare- to sprout
-ion state, process, or quality of
To begin or cause to sprout or grow.

Germovitellarium
Latin
germen- a bud, offshoot
-vitellus- yolk
-ium quality or relationship
Closely associated ovary and yolk-producing structures in rotifers.

Gestation
Latin
gestare- to bear
-ion state, process, or quality of
Time during which a placental mammal develops in a uterus.

Getter
Middle English
geta- to obtain
-er one that performs an action
A chemically active substance such as magnesium that is ignited in vacuum tubes to remove traces of gas, or any substance that is added to another to remove traces of impurities.

Geyser
Icelandic
geysa to gush
A natural hot spring that intermittently ejects a column of water and steam into the air.

Gibbous
Latin
gibbus bulging, hunch-backed, humped
Pertaining to swelling by a regular curve or surface; protuberant; convex, as "the moon is gibbous between the half moon and the full moon."

Gizzard
Latin
gigeria giblet, cooked entrails of poultry
The thickened part of the alimentary canal in some animals (such as an insect or earthworm) that is similar to the crop of a bird.

Glabrate
Latin
glab- smooth or hairless
-ate of or having to do with
Becoming smooth or glabrous from age.

Glacial
Latin
glacialis ice
Having an icelike form in its pure state at or just below room temperature.

Gland
Latin
glans acorn
A term applied to a group of organs that secrete chemicals used in other parts of the body.

Glaucoma
Greek
glaukos- gray
-oma swelling
A disease of the eye caused by increased pressure, which can damage the optic nerve and result in blindness.

Glitch
Yiddish/German
glitschn lapse, slip
A sudden change in the period of rotation of a neutron star.

Globular
Latin
globus- globular mass
-ar relating to or resembling
In biology, globe-shaped, having the form of a ball or sphere (e.g., globular proteins)

Globular cluster
Latin/Old English
globus- globular mass
-ar relating to or resembling
clyster bunches
In astronomy, a system of stars, generally smaller in size than a galaxy, that is more or less globular in conformation.

Glochidium
Greek
glokhis- point, barb of an arrow
-idion quality of relationship
Bivalved larval stage of freshwater mussels.

Glomerulus
Latin
glomer- ball
-ulus of, relating to, or resembling
Capillary network within glomerular capsule.

Glossus
Greek
glw^ssa the tongue
The muscular organ found in the mouths of vertebrates. It is involved with the manipulation of food during chewing, tasting, and swallowing, and with speech.

Glottis
Greek
glotta/glossa tongue
The opening between the vocal cords in the larynx.

Glucagon
Greek
glukus- sweet, sweetness
-agein lead, drive
A peptide hormone secreted by pancreatic endocrine cells that raises blood glucose levels; an antagonistic hormone to insulin.

Glucolytic
Greek
glukus- sweet, sweetness
-ly- (luein) to loosen, dissolve, dissolution, break
-ic (ikos) relating to or having some characteristic of
Pertaining to the metabolic breaking down of glucose for the production of ATP occurring in the cytoplasm of cells.

Gluon
Latin
gluton- glue
-on subatomic particle
A hypothetical, massless, neutral elementary particle believed to mediate the strong interaction that binds quarks together.

Glycogen
Greek
glukus- sweet, sweetness
-gen to give birth, kind, produce
A polysaccharide that is the main form of carbohydrate storage in animals and occurs primarily in the liver and muscle tissue. It is readily converted to glucose as needed by the body to satisfy its energy needs. Also called animal starch.

Glycolysis
Greek
glykys- sweet
-ly- (*luein*) to loosen, dissolve, dissolution, break
-sis action, process, state, condition
Initial reactions of both aerobic and anaerobic pathways by which glucose is partially broken down to pyruvate, with a net yield of two ATP. Glycolysis proceeds in the cytoplasm of all cells, and oxygen has no role in it.

Gnathostomes
Greek
gnathos- jaw
-stoma mouth
The group of vertebrates with distinct jaws.

Gonad
Greek
gonos procreation, genitals
A reproductive organ that produces sperm or eggs.

Gonadotropin
gonos- procreation, genitals
-trope- bend, curve, turn, a turning; response to stimulus
-in protein or derived from a protein
Any one of three hormones released by either the pituitary gland or the placenta. These hormones stimulate the gonads and control reproductive activity.

Gonangium
Latin
gonos- seed, procreation
-angeion diminutive of *vessel*
Reproductive zooid of hydroid colony (Cnidaria).

Gonophore
Latin
gonos- seed, procreation
-pherein to carry

A small reproductive organ found in some sponges.

Gonopore
Greek
gonos- seed, procreation
-poros an opening
A genital pore found in many invertebrates.

Gradation
Latin
gradus- walk, step, take steps, move around
-ion state, process, or quality of
The leveling of a planet's surface through weathering, erosion, transpiration, and deposition of rock debris by water, wind, and gravity.

Gradient
French (from Latin)
grade- a position in a scale of size, quality, or intensity
-ient performing, promoting, or causing a specific action
The rate at which a physical quantity changes with respect to a given variable.

Gradualism
Latin
gradus- walk, step, take steps, move around
-ism state or condition, quality
The evolution of new species by the slow, steady accumulation of small genetic changes occurring over long periods of time.

Granuloma
Latin
granum- grain, seed
-oma community
A mass of inflamed granulation tissue, usually associated with ulcerated infections.

Granum
Latin
granum grain, seed
A stacked, membranous structure within a chloroplast that contains the chlorophyll and is the site of the light reactions involved in photosynthesis.

Gravitropism
Latin
gravis- heavy, weighty
-trope- bend, curve, turn, a turning; response to stimulus
-ism state or condition, quality
A turning or growth movement by a plant in response to gravity.

Gravity
Latin
gravis- heavy, weighty
-ity state of, quality of
An acceleration value related to the force attracting two bodies.

Guanine
Spanish
huanu- the dung of sea birds or bats
-ine of or relating to
A purine base, $C_5H_5ON_5$, that is an essential constituent of both RNA and DNA.

Gully
French
goulet the throat
Erosional features; deep channels found in sedimentary layers, acted on by weathering.

Gustation
Latin
gustare- to taste
-ion state, process, or quality of
The sense of taste; the ability or the act of tasting.

Guttation
Latin
gutta- to drop
-ion state, process, or quality of
The exudation of water from leaves resulting from root pressure.

Gymnosperm
Greek
gumnos- naked
-sperma seed
A plant whose seeds are not enclosed within an ovary.

Gynecophoric
Greek
gyne- woman, female
-pherein to carry
Pertains to the groove in male schistosomes (certain trematodes) that carries the female.

Gynenosia
Greek
gyne- woman, female
-nosia disease
A disease occurring most often in females.

Gynoecium
Greek
gyne- woman, female
-oikos- house
-ium quality or relationship
Part of a flower that houses the female gametophytes, the pistils.

Gyroscope
Greek
gyros- ring, compass
-skopion for viewing with the eye
Rotating mechanism in the form of a universally mounted spinning wheel that offers resistance to turns in any direction.

H

Habitat
Latin
habitare to dwell
Area or environment where an organism or ecological community normally lives.

Hadean
Greek
haides mythological subterranean world of the departed spirits
Relates to the beginning of the earth's formation, when the surface was molten and forming, 4.5–3.8 billion years ago (bya).

Hadron
English (from Greek)
hadros- thick
-on a particle
Any of a class of subatomic particles that are composed of quarks and take part in the strong interaction.

Halic
Greek
hal- salt
-ic (ikos) relating to or having some characteristic of
Pertaining to saline or saltlike conditions.

Halimetry
Greek
hal- salt
-metria (metron) the process of measuring
The measurement of the amount of saline matter in solution.

Halite
Greek
hal- salt
-ite minerals and fossils
A colorless, crystalline rock salt found in salt marshes, dried desert floors, and mines.

Halobiotic
Greek
hal- salt
-bios- life, living organisms, or tissue
-ic (ikos) relating to or having some characteristic of
Refers to life in the sea, to organisms capable of living in a marine environment.

Halogen
Greek
halos- disk of sun
-gen to give birth, kind, produce
Reactive, nonmetallic element in group 7A of the periodic table.

Halolimnetic
Greek
hal- salt
-limn- lake
-ic (ikos) relating to or having some characteristic of
Pertaining to salt lakes; marine organism designed to live in freshwater.

Halopexia
Greek
hal- salt
-pexia attaching to or fixation
The physiological retention of salt by the body.

Halophile

Greek

hal- salt

-phile one who loves or has a strong affinity or preference for

A microorganism requiring a high concentration of salt for optimal growth.

Halophobe

Greek

hal- salt

-phobos fear

Any creature that is intolerant of saline life.

Harmonics

Greek

harmonikos- harmony

-ic (ikos) relating to or having some characteristic of

Tones whose frequencies are whole-number multiples of the fundamental; also referred to as fundamental frequencies.

Haustoria

Latin

haurire- to drink

-ia names of diseases, place names, or Latinizing plurals

The hyphae that invade the cells of a host to absorb nutrients.

Heat

Old English

hete hot

A form of energy associated with the motion of atoms or molecules.

Helictite

Greek

helix- spiral

-ite a part of or product of

Thin crystal strains that resemble flowers and are found in clusters on cave ceilings.

Heliocentric

Greek

helio- sun

-kentron- a point or place that is equally distant from the sides or outer boundaries of something; the middle

-ic (ikos) relating to or having some characteristic of

Describes the nature of the solar system, with the sun located in the center and the planets orbiting around it.

Hematemesis

New Latin

haimat- blood

-emesis vomit

The presence of blood or blood cells in vomit.

Isaac Newton

Beginning in 1665 and continuing into 1666, the Great Plague of London devastated the English population. This catastrophic disease, most likely bubonic plague, killed over 75,000 in that country. Because of these conditions, a relatively young undergraduate student at Cambridge University in London was sent home. At Woolthorpe, the town where he was born, Isaac Newton would live as a recluse during that year, far from the death and dying in London.

With the exception of Einstein's miracle year of 1905, few other single years in history have had such a dramatic impact on science, discovery, and the progression of thought. In the 18 months during his time off from school, Isaac Newton laid some of the groundwork for the study of optics and the nature of light, he invented calculus, and he put forth some of the essential elements for his theory of universal gravitation.

Isaac Newton was another major figure of the scientific revolution. Like most other great thinkers of his day, he was, for a time, fascinated by mysticism, astrology, and mathematics. He sought harmony in the universe through mathematics.

Among Newton's theories was the idea that gravity is universal. He postulated that if the earth's gravitational attraction held the moon in its orbit, then this same force was responsible for keeping other planets in their orbits as well. The orbital paths of planets were affected, in part, by the gravitational attraction of the sun. Newton, unlike Kepler, was able to mathematically prove Kepler's laws of planetary motion.

Isaac Newton is known for his three laws of motion.

- Newton's first law, the law of inertia, states that an object at rest tends to stay at rest and that an object in motion tends to stay in motion unless acted upon by a net external force.

- Newton's second law states that force = mass × acceleration. That is, the acceleration produced by a net force on an object is directly proportional to the magnitude of the net force and is inversely proportional to the mass.
- Newton's third law states that for every action there is an equal and opposite reaction.

On July 5, 1687, Isaac Newton published his seminal three-volume work, *Philosophiae Naturalis Principia Mathematica,* which is Latin for *Mathematical Principles of Natural Philosophy.* His text is sometimes referred to as *Principia* or *Principia Mathematica.* It contains his groundbreaking principles for the mechanics of the universe, his three laws of motion, and his law of universal gravitation.

Sir Isaac Newton died on March 20, 1727, in London.

Hematocrit
Greek
haimat- blood
-krites judge
The instrument used to determine the ratio of the volume occupied by blood cells to the total volume of blood.

Hematolysis (hemolysis)
Greek
haimat- blood
-ly- (luein) to loosen, dissolve; dissolution, break
-sis action, process, state, condition
The lysing or breakdown of erythrocytes (red blood cells) with the subsequent release of hemoglobin.

Hematuria
New Latin
haimat- blood
-uria urine
The presence of blood or blood cells in urine

Hemimetabolous
Greek
hemi- half
-metabole- change
-ous full of, having the quality of, relating to
Refers to gradual metamorphosis during the development of insects, without a pupal stage.

Hemiptera
Greek

hemi- half
-pteron wing
Insect order for true bugs; wingless or four-winged bugs that include such insects as bedbugs and chinch bugs.

Hemisphere
Greek
hemi- half
-sphaira a globe shape, ball, sphere
A half of a sphere.

Hemocoel
Greek
haima- blood
-koilos cavity
A cavity or series of spaces between the organs of most arthropods and mollusks through which blood circulates.

Hemodialysis
Greek
haimo- relating to blood or blood vessels
-dia- through, across, apart
-ly- (luein) to loosen, dissolve; dissolution, break
-sis action, process, state, condition
A medical procedure for removing metabolic waste products from the blood.

Hemoglobin
Latin/Greek
haimo- relating to blood or blood vessels
-globulus- globule
-in protein or derived from protein
An iron-containing respiratory pigment occurring in vertebrate red blood cells and in blood plasma of many invertebrates; a compound of an iron porphyrin heme and a protein globin.

Hemolymph
Latin/Greek
haimo- relating to blood or blood vessels
-numphe clear fluid; water nymph, young bride
Fluid in the coelom or hemocoel of some invertebrates that represents the blood and lymph of vertebrates.

Hemolysis (hematolysis)
Greek
haimo- relating to blood or blood vessels
-ly- (luein) to loosen, dissolve; dissolution, break
-sis action, process, state, condition
The destruction of red blood cells, leading to the release of hemoglobin from the cells into the blood plasma.

Hemophilia
Greek
haimo- relating to blood or blood vessels
-phile- one who loves or has a strong affinity or preference for

-ia names of diseases, place names, or Latinizing plurals

A group of hereditary bleeding disorders characterized by a deficiency of one of the factors necessary for coagulation of the blood.

Hemorrhage
Greek
haimo- relating to blood or blood vessels
-rhegnynai to break, burst
Excessive discharge of blood from the blood vessels; profuse bleeding from a ruptured blood vessel.

Hemorrhoid
Greek
haimo- relating to blood or blood vessels
-rhein- to flow
-oid (oeides) resembling, having the appearance of
A mass of dilated blood vessels located in the anus; the dilated vessels cause pain and itching.

Hepatitis
Latin
hepat- liver
-itis inflammation, burning sensation
A disease or condition marked by inflammation of the liver.

Hepatomalacia
Greek
hepat- liver
-malacia softening of tissue
A disease or condition of the liver marked by distinct softening of the fleshy tissue of the liver.

Hepatonecrosis
Greek
hepta- liver
-necr- death
-sis action, process, state, condition
Death of liver cells, usually caused by either a pathogenic organism or a toxic substance.

Hepatorrhexis
Greek
hepta- liver
-orrhexis, -rrhexis rupture of an organ or vessel; a breaking forth, bursting
The rupturing of the liver occurring as a result of injury or disease.

Heptad
Greek
heptados group of seven
An element, atom, or radical that has a valence of 7.

Herbicide
Latin

herba- grass, green crops
-cide (caedere) to cut, kill, hack at, or strike
Any chemical agent that is toxic to some or all plants and is used to destroy unwanted vegetation.

Herbivore
Latin
herba- grass, green crops
-vorare to devour
Any organism subsisting on plants.

Heredity
Latin
hered- heir
-ity state of, quality of
The transmission of qualities from ancestor to descendant through the genes.

Hermaphrodite
Greek
hermes- Hermes, Greek god of boundaries
-aphrodite Aphrodite, Greek goddess of love and beauty
An animal or plant species that normally exhibits both male and female sex organs.

Hernia
Latin
herni- protruded viscus; rupture
-ia names of diseases, place names, or Latinizing plurals
The protrusion of a bodily organ through a normally intact supporting wall-like structure.

Heterocercal
Greek
heteros- different
-kerkos- tail
-al of the kind of, pertaining to, having the form or character of
In some fish, having or referring to a tail with the upper lobe larger than the lower, and the end of the vertebral column somewhat upturned in the upper lobe, as in sharks.

Heterochrony
Greek
heteros- different
-khronos- time
-y place for an activity; condition, state
Evolutionary change in the relative time of appearance or rate of development of characteristics from ancestor to descendant.

Heterocyst
Greek
heteros- different
-cyst (kustis) sac or bladder containing fluid

A large, thick-walled, transparent cell that occurs at intervals along the filaments of certain cyanobacteria.

Heterodont
Greek
heteros- different
-odous tooth
Having teeth differentiated into incisors, canines, and molars for different purposes.

Heterotroph
Greek
heteros- different
-trophos (trophein) to nourish, food; nutrition; development
An organism that obtains both organic and inorganic raw material from its environment in order to survive.

Heterozygote
Greek
heteros- different
-zygoun to yoke
An organism that has different alleles at a particular gene locus on homologous chromosomes.

Hexabasic
Latin
hexa- six
-bas- low
-ic (ikos) relating to or having some characteristic of
Relates to having six hydrogen atoms that can be replaced by basic atoms or radicals.

Hexactinellida
Greek
hexa- six
-aktin- ray
-ella little
A siliceous sponge characterized by glassy spicules.

Hexagonal
Greek
hexa- six
-agon- a violent, intense struggle
-al of the kind of, pertaining to, having the form or character of
Having three equal axes intersecting at angles of 60 degrees in one plane, and one axis of variable length that is perpendicular to the others.

Hexahedron
Greek
hexa- six
-hedron face
A Platonic six-sided solid; a cube.

Hexamerous
Greek

hexa- six
-meros part
Having six parts; specifically, symmetry based on six or multiples thereof.

Hibernation
Latin
hibern- winter
-ation state, process, or quality of
The process of spending the winter in a resting state.

Hilum
Latin
hilum trifle
A notch on the medial surface of the kidney where blood vessels enter and leave the kidney.

Hippocampus
Latin
hippos- riverine
-kampos sea monster
Composed of gray matter, this ridge on the floor of the lateral ventricles of the brain is responsible for memory.

Hippopotamus
Greek
hippos- riverine
-potamios horse
Chiefly aquatic mammal with an extremely large head and mouth, bare and very thick grayish skin, and short legs.

Histochemistry
Greek
histos- web, tissue
-chemo- (khemeia) chemical; alchemy
-metria (metron) the process of measuring
The science dealing with the chemical composition of the tissues of the body.

Histology
Greek
histos- web, tissue
-logy (logos) used in the names of sciences or bodies of knowledge
The study of the microscopic structures of tissues.

Histone
Greek
histos- web, tissue
-one chemical compound containing oxygen in a carbonyl group
Any of a group of strongly basic low-molecular-weight proteins that combine with nucleic acid to form nucleoproteins.

Holeuryhaline
Greek
holos- complete, whole, entire, all, full

-eury- wide
-hal- salt
-ine in a chemical substance
Refers to organisms that freely inhabit freshwater, sea water, and brackish water.

Holistic
Greek
holos- complete, whole, entire, all, full
-ist- one who performs an action
-ic (ikos) relating to or having some characteristic of
Describes an approach to medical care that emphasizes the study of all aspects of a person's health, including physical, psychological, social, economic, and cultural factors.

Holocene
Greek
holos- complete, whole, entire, all, full
-kainos recent
An epoch of the Quaternary period, spanning the time from the end of the Pleistocene to the present.

Holoenzyme
Latin
holos- complete, whole, entire, all, full
-en- in, at, onto
-zume ferment, leaven
A fully active, complex enzyme, composed of a protein and a coenzyme.

Holometabolous
Greek
holos- complete, whole, entire, all, full
-meta- between, after, beyond, later
-bol- (ballein) to put or throw
-ous full of, having the quality of, relating to
Pertains to complete metamorphosis during development.

Holophytic
Greek
holos- complete, whole, entire, all, full
-phyt- plant
-ic (ikos) relating to or having some characteristic of
Relates to the process that occurs in green plants and certain protozoa involving synthesis of carbohydrates from carbon dioxide and water in the presence of light, chlorophyll, and certain enzymes.

Holozoic
Greek
holos- complete, whole, entire, all, full
-zoikos- of animals
-ic (ikos) relating to or having some characteristic of
Describes a type of nutrition involving ingestion of liquid or solid organic food particles.

Homeopathy
Greek
homeo- same, like, resembling, sharing, similar, equal
-pathos- feeling, sensation, perception
-y place for an activity, condition, state
A method of disease treatment that involves the administration of small doses of chemicals that, if given in large amounts, would produce symptoms in healthy people that are similar to those found in people with the disease.

Homeostasis
Greek
homeo- same, like, resembling, sharing, similar, equal
-statos- standing, stay, make firm, fixed, balanced
-sis action, process, state, condition
Tendency of an organism to maintain internal equilibrium of temperature and fluid content, for example, by regulation of its bodily processes.

Homeothermic
Greek
homeo- same, like, resembling, sharing, similar, equal
-thermos- combining form of "hot" (heat)
-ic (ikos) relating to or having some characteristic of
Having a nearly uniform body temperature.

Hominid
Latin
homo/homonis- man
-id state, condition; having, being, pertaining to, tending to, inclined to
A member of the family Hominidae; human beings are the only surviving species.

Homocercal
Greek
(h)omos- (combining form) one and the same, common
-kerkos tail
Having or referring to a tail with the upper and lower lobes symmetrical and the vertebral column ending near the middle of the base, as in most teleost fish.

Homogeneous
Greek
(h)omos- (combining form) one and the same, common
-genus offspring, kind
Of the same or similar nature or kind.

Homologous
Greek
(h)omos- (combining form) one and the same, common
-logos word, proportion
Having the same or similar proportions or characteristics. In genetics, having the same gene sequence on two different chromosomes.

Homoplasy
Greek
(h)omos- (combining form) one and the same, common
-plasy growth or development of
Independent evolution of similar or identical characteristics through convergence or parallel evolution.

Homozygote
Greek
(h)omos- (combining form) one and the same, common
-zugoun to yoke
Organism having the two genes at corresponding loci on homologous chromosomes identical for one or more loci.

Horizontal
Greek
horos- (horizein) to limit; boundary
-al of the kind of, pertaining to, having the form or character of
Refers to the axis parallel to the horizon (side by side); of or near the horizon; relating to the horizon.

Hormone
Greek
horman that which sets in motion; to urge on
Substances produced by a gland or tissue, then transported by the blood to effect physiological activity and regulate development.

Horology
Greek
horo- hour, period of time, season, time
-logy (logos) used in the names of sciences or bodies of knowledge
The science of measuring time.

Horoscope
Greek
horo- hour, period of time, season, time
-skopos observer
An astrological prediction based on observations of the positions of celestial objects.

Horse
Old English
hors horse

Common name given to species of the genus *Equus*. These mammals are characterized by having long legs, short-haired coats, long tails, and hooved feet.

Humerus
Latin
humer- shoulder, upper arm
-us thing
The long bone of the arm or forelimb, extending from the shoulder to the elbow.

Humidity
Latin
humidus- moist, wet
-ity state of, quality of
The amount of water vapor or moisture in the air.

Humoral
Middle English
humor- fluid
-al of the kind of, pertaining to, having the form or character of
Of or pertaining to the fluid of a body.

Humus
Latin
humus soil
Partially decomposed organic matter consisting of both plant and animal remains, rich in nutrients and capable of holding significant amounts of water.

Hyaline
Greek
hualos- glass
-in protein or derived from a protein
A clear, homogeneous, glassy substance normally found in cartilage, vitreous humor, mucin, and glycogen, and pathologically found in the degeneration of tissues and cells.

Hybrid
Latin
hybrida mongrel
An offspring of two animals or plants of different races, breeds, varieties, species, or genera.

Hybridization
Latin
hybrida- mongrel
-ation action, process, state, or condition
The act of cross-breeding various species or subspecies of organisms.

Hydra
Greek
hydra of or having to do with water
In astronomy, the largest constellation, winding across more than a quarter of the sky.

Hydranth
Greek
hydr- of or having to do with water
-anthos flower
Nutritive zooid of hydroid colony.

Hydrate
Greek
hydr- of or having to do with water
-ate of or having to do with
A compound that contains a specific ratio of water to ionic compound.

Hydration
Greek
hydr- of or having to do with water
-ion state, process, or quality of
In chemistry, the combination of water and another substance to obtain a single product. In earth science, a form of chemical weathering caused by the expansion of certain minerals as they absorb water.

Hydraulic
Greek
hydr- of or having to do with water
-aulos characterized by having a hollow way; tube, pipe
Of or relating to water or other liquid in motion.

Hydrocarbon
Greek
hydr- of or having to do with water
-carbon coal, charcoal
Organic compounds containing hydrogen and carbon only.

Hydrocephalus
Greek
hydr- of or having to do with water
-cephalo- (kephalikos) head
-us thing
A usually congenital condition in which an abnormal accumulation of fluid in the cerebral ventricles causes enlargement of the skull and compression of the brain.

Hydrocoel
Greek
hydr- of or having to do with water
-koilos hollow
Second or middle coelomic compartment in echinoderms; the left hydrocoel gives rise to the water-vascular system.

Hydrocoral
Greek
hydr- of or having to do with water
-korallion coral
Any of certain members of the cnidarian class Hydrozoa that secrete calcium carbonate and resemble true corals.

Hydroformylation
Greek/Middle English
hydr- of or having to do with water
-formyl- the negative univalent radical HCO
-ion state, process, or quality of
The process by which an –H and a –CHO are added across a carbon-carbon double bond. An aldehyde synthesis process.

Hydrogenation
Greek
hydr- of or having to do with water
-gen- to give birth, kind, produce
-ation state, process, or quality of
The process of combining a substance with hydrogen.

Hydrogeology
Greek
hydr- of or having to do with water
-ge- earth
-logy (logos) used in the names of sciences or bodies of knowledge
The branch of geology that deals with the occurrence, distribution, and effects of groundwater.

Hydrology
Greek
hydr- of or having to do with water
-logy (logos) used in the names of sciences or bodies of knowledge
The study of the properties, distribution, and effects of water on the surface of the earth, the atmosphere, and the earth's substrate.

Hydrolysis
Greek
hydr- of or having to do with water
-ly- (luein) to loosen, dissolve; dissolution, break
-sis action, process, state, condition
Decomposition of a chemical compound by reaction with water, such as the dissociation of a dissolved salt or the catalytic conversion of starch to glucose.

Hydrometer
Greek
hydr- of or having to do with water
-meter (metron) instrument or means of measuring; to measure
An instrument used to determine specific gravity.

Hydropenia
Greek
hydr- of or having to do with water
-penia reduction, poverty, lack, deficiency
A condition or disorder that results in a reduction of water.

Hydrophobic
Greek/Latin
hydr- of or having to do with water
-phob- fear, lacking an affinity for
-ic (ikos) relating to or having some characteristic of
Describes something that is repelled by water or tends not to combine with or dissolve in water.

Hydrophyte
Greek
hydr- of or having to do with water
-phyte plant
A plant adapted to grow in water; a water lily.

Hydroplane
Greek
hydr- of or having to do with water
-plane surface
To skim along the surface of water.

Hydroponic
Greek
hydr- of or having to do with water
-pono- work
-ic (ikos) relating to or having some characteristic of
Pertains to growing plants without soil in nutrient-enriched water.

Hydropower
Greek/Latin
hydr- of or having to do with water
-potis able, powerful
Electrical energy produced by falling or flowing water.

Hydrosphere
Greek
hydr- of or having to do with water
-sphaira a globe shape, ball, sphere
The water on the earth's surface.

Hydrostatic
Greek
hydr- of or having to do with water
-statos- standing, stay, make firm, fixed, balanced
-ic (ikos) relating to or having some characteristic of
Relating to fluids at rest or to the pressures they exert or transmit.

Hydrothermal
Greek
hydr- of or having to do with water
-thermos- combining form of "hot" (heat)
Relating to hot water; magmatic releases are rich in water.

Hydrozoan
Greek
hydr- of or having to do with water
-zoon animal, animal-like

Any of a group of freshwater coelenterates including hydras, hydroids, hydrocorals, and siphonophores.

Hygrometer
Greek
hygr- wet or moist
-meter (metron) instrument or means of measuring; to measure
An instrument that measures humidity.

Hygroscopic
Greek
hygr- wet, moist
-scopion- to look at, examine
-ic (ikos) relating to or having some characteristic of
Refers to a substance that easily absorbs water from the air to become a hydrate.

Hymen
Greek
humen thin skin, membrane
A membranous tissue fold that either partially or completely covers the vaginal orafice.

Hymenoptera
Greek
humen- thin skin, membrane
-pteron wing
Order of insects characterized by thin, membranous wings. Most have two pairs of wings, with the first being considerably larger than the second. Includes wasps, bees, and ants.

Hyoid
Greek
hu- upsilon, Greek letter *U*
-oid (oeides) resembling, having the appearance of
Relating to the hyoid bone.

Hyperglycemia
Greek
hyper- above, high
-glyco- sugar
-emia the condition of having (a specific thing) in the blood
Abnormally high blood sugar.

Hyperpnea
Greek
hyper- over, beyond
-pnein breathing or breath
Abnormally deep or rapid breathing.

Hypertension
Greek
hyper- over, beyond
-tens- stretching; physiological imbalance
-ion state, process, or quality of
Abnormally high blood pressure.

Hyperthermic
Greek
hyper- over, beyond
-thermos- combining form of "hot" (heat)
-ic (ikos) relating to or having some characteristic of
Having the characteristics of or relating to a condition of unusually high body temperature.

Hypertonic
Greek
hyper- over, beyond
-ton- tension
-ic (ikos) relating to or having some characteristic of
Having the higher osmotic pressure of two solutions.

Hyperventilation
Greek
hyper- over, beyond
-ventilare- to fan
-ion state, process, or quality of
A pulmonary ventilation rate that is higher than what is necessary for normal pulmonary gas exchange.

Hyphae
Greek
huphe web
Threadlike filaments found in the mycelium of a fungus.

Hypocalcemia
Greek/Latin
hypo- under, below, beneath, less than, too little, deficient
-calc- calcium
-emia the condition of having a (specific thing) in the blood
A deficiency of calcium in the blood.

Hypochondria
Greek
hypo- under, below, beneath, less than, too little, deficient
-khondr- grain, any small rounded mass; cartilage, gristle, granule, or a relationship to cartilage
-ia names of diseases, place names, or Latinizing plurals
A disorder characterized by a misinterpretation of physical signs that leads to the belief that one has a serious disease even though repeated evaluations show no indications of any physical disorder.

Hypodermis
Greek/Latin
hypo- under, below, beneath, less than, too little, deficient
-derma skin
The cellular layer lying beneath and secreting the cuticle of annelids, arthropods, and certain other invertebrates.

Hypoglossal
Greek
hypo- under, below, beneath, less than, too little, deficient
-gloss- tongue
-al of the kind of, pertaining to, having the form or character of
Of or relating to the area under the tongue.

Hypognathous
Greek
hypo- under, below, beneath, less than, too little, deficient
-gnathos jaw
Pertains to having the head directed vertically and the mouthparts directed ventrally.

Hypokalemia
Greek
hypo- under, below, beneath, less than, too little, deficient
-kali- potassium
-emia the condition of having (a specific thing) in the blood
A deficiency of potassium in the blood.

Hypostome
Greek
hypo- under, below, beneath, less than, too little, deficient
-stoma mouth
Name applied to the structure in various invertebrates, such as mites and ticks, that is located at the posterior or ventral area of the mouth; elevation supporting the mouth of a hydrozoan.

Hypotenuse
Greek
hypo- under, below, beneath, less than, too little, deficient
-teinein to stretch
The line segment stretched under the right angle; the line opposite the right angle in a right triangle.

Hypothalamus
Greek
hypo- under, below, beneath, less than, too little, deficient
-thalamos inner chamber, bedroom
The region of the brain situated below the thalamus and above the pituitary gland, which acts as a control center for the autonomic nervous system and for hormonal activity.

Hypothermia
Greek
hypo- under, below, beneath, less than, too little, deficient
-thermos- combining form of "hot" (heat)

-ia names of diseases, place names, or Latinizing plurals
A condition in homeothermal organisms marked by a drop to a temperature below normal.

Hypothesis
Greek
hypo- under, below, beneath, less than, too little, deficient
-tithenai- to put or place
-sis action, process, state, condition
An assertion made as a possible explanation for a problem.

Hypothetical
Greek
hypo- under, below, beneath, less than, too little, deficient
-tithenai- to put or place
-alis of, related to
Refers to a situation or setting based on or relating to a hypothesis.

Hypotonic
Latin/Greek
hypo- under, below, beneath, less than, too little, deficient
-ton- tension
-ic (ikos) relating to or having some characteristic of
In chemistry, refers to a situation where one solution's osmotic pressure is lower than that of another solution.

Hypoxia
Greek
hypo- under, below, beneath, less than, too little, deficient
-ox- acid, acidic
-ia names of diseases, place names, or Latinizing plurals
A disorder that causes a reduction in the oxygen supply to tissues.

Hysterectomy
Greek
hustera- uterus, womb
-ekt- outside, external, beyond
-tomos (temnein) to cut, incise, section
Partial or complete surgical removal of the uterus.

Hysteroptosis
Greek
hyster- the womb or uterus; hysteria
-pto- fall, a falling down of an organ; drooping, sagging; corpse
-sis action, process, state, condition
The sagging or prolapsing of the female uterus.

Hystolytic
Greek
histos- web, tissue
-ly- (luein) to loosen, dissolve; dissolution, break
-ic (ikos) relating to or having some characteristic of
Pertaining to the degeneration of tissues.

Ichthyologist

Greek

ichthus- fish

-ologist one who deals with a specific topic

A scientist who studies the biology of fish.

Ichthyology

Greek

ichthus- fish

-logy (logos) used in the names of sciences or bodies of knowledge

Branch of zoology that deals with the study of fish.

Icosahedron

Greek

icosa- twenty

-hedron face

A Platonic solid with twenty faces.

Ideal

Latin

idea- a plan, scheme, notion, or method

-al of the kind of, pertaining to, having the form or character of

Conforming to an ultimate form or standard of perfection or excellence.

Igneous

Latin

ignis- fire

-ous full of, having the quality of, relating to

Refers to molten rock that cools and solidifies.

Ileum

Latin

ileum groin, flank

The terminal end of the small intestine; it extends from the jejunum to the ileocecal sphincter.

Iliocostal

Latin

ilia- groin, flank

-costo- rib

-al of the kind of, pertaining to, having the form or character of

Relating to the ilium and ribs.

Image

Latin

imago image

In optics, the likeness of an object produced by the use of a lens or group of lenses.

Imbibition

Latin

in- in, into, toward, against, on, upon

-bib- drink

-ion state, process, or quality of

Adsorption of water to internal surfaces of an organism, leading to swelling.

Immigrate

Latin

in- in, into, toward, against, on, upon

-migrare- to go into, to depart

-ion state, process, or quality of

To enter and settle in a country or region to which one is not native.

Immiscible

Latin

in- in, into, toward, against, on, upon
-miscere- to mix
-ible capable
Refers to that which cannot undergo mixing or blending.

Immunotherapy
Latin/Greek
immunis- not affected by a given influence; unresponsive
-therapeuein to treat medically
Treatment of disease by inducing, enhancing, or suppressing an immune response.

Impedance
Latin
impedire to hinder motion on foot
A measure of the total opposition to current flow in an alternating current circuit, made up of two components: ohmic resistance and reactance.

Impenetrability
Latin
im- not
-penitus- deeply, permeate
-ity state of, quality of
A property of matter where no two objects can occupy the same space at the same time.

Impulse
Latin
impellere to impel
The product obtained by multiplying the average value of a force by the time during which it acts. The impulse equals the change in momentum produced by the force during this time interval.

Inactive
Latin
in- in, into, toward, against, on, upon
-agere to drive or do
Not active; in biology, refers to a condition during which metabolism is marked by a reduction of activity, possibly because of an infection.

Incandesce
Latin
in- in, into, toward, against, on, upon
-candescere become white hot
To glow or cause to glow with heat.

Incisor
Latin
in- in, into, toward, against, on, upon
-caedere- to cut
-or a condition or property of things or persons; person who does something
A tooth for cutting or gnawing, located at the front of the mouth in both jaws.

Incline
Latin
in- in, into, toward, against, on, upon
-klinein to lean, sloping
A slant; deviation from the horizontal or vertical.

Incubation
Latin
in- in, into, toward, against, on , upon
-cubare- to lie down on
-ion state, process, or quality of
Maintenance of optimal conditions for growth and development.

Indigenous
Latin
in- in, into, toward, against, on, upon
-genus- birth, origin, kind
-ous full of, having the quality of, relating to
Pertaining to a group of organisms native and original to a region.

Induction
Latin
in- in, into, toward, against, on, upon
-ducere- to lead
-tion action, process or quality of
The production of magnetism or electromotive force, or the separation of charge from a body by a neighboring body not in contact with it.

Inductor
Latin
in- in, into, toward, against, on, upon
-ducere- to lead
-or a condition or property of things or persons
A coil of wire that generates a magnetic field when a current is passed through it.

Inelastic
Greek
in- in, into, toward, against, on, upon
-elaunein- to beat out
-ic (ikos) relating to or having some characteristic of
Refers to a type of collision in which two objects remain attached after the collision.

Inert
Latin
in- in, into, toward, against, on, upon
-aras skill
Unable to move or act; not readily reactive with other elements.

Inertia
Latin
iners- idleness
-ia names of diseases, place names, or Latinizing plurals

The tendency of a body to resist acceleration; the tendency of a body at rest to remain at rest, or of a body in straight-line motion to stay in motion in a straight line unless acted on by an outside force.

Infectious
Latin
in- in, into, toward, against, on, upon
-facere- to make, do, build, cause, produce; forming, shaping
-ous full of, having the quality of, relating to
Pertaining to a contagious disease capable of spreading rapidly to others.

Inference
Latin
in- in, into, toward, against, on, upon
-ferre- to bear
-ence the condition of
The act of passing from one proposition, statement, or judgment considered true to another, whose truth is believed to follow from that of the former.

Inferno
Latin
infernus hell, lower, underground
In astrophysics, a unit for describing the temperature inside a star. One inferno is approximately one billion degrees celsius.

Inflammation
Latin
in- in, into, toward, against, on, upon
-flamma- flame
-ation action, process, or quality of
A localized defensive reaction of body tissue to irritation, damage, or infection; characterized by pain, redness, swelling, and sometimes loss of function.

Inflation
Latin
in- in, into, toward, against, on, upon
-flare- to blow
-ion state, process, or quality of
In astronomy, an extremely brief phase of ultra-rapid expansion of the very early universe.

Influenza
Latin
in- in, into, toward, against, on, upon
-fluere- to flow, wave
-za quality or state
A human respiratory infection of undetermined cause.

Infraciliature
Latin
infra- inferior to, below, or beneath
-cilia- eyelashes

-ure act, process, condition
The organelles just below the cilia in ciliate protozoa.

Infracostal
Latin
infra- inferior to, below, or beneath
-costo- rib
-al of the kind of, pertaining to, having the form or character of
Pertaining to or referring to a region below the ribs.

Infrasonic
Latin
infra- inferior to, below, or beneath
-sonus- sound
-ic (ikos) relating to or having some characteristic of
Generating or using waves or vibrations in frequencies below that of audible sound.

Inherit
Latin
in- in, into, toward, against, on, upon
-hereditare to inherit
To acquire or express traits or conditions through transmission of genetic material from parents to offspring.

Initiator
Latin
initium- beginning
-or a condition or property of things or persons
A substance or chemical that begins a reaction but is consumed or chemically changed in the reaction.

Inorganic
Latin
in- in, into, toward, against, on, upon
-organon- instrument
-ic (ikos) relating to or having some characteristic of
Composed of nonliving matter.

Insect
Greek
in- in, into, toward, against, on, upon
-secare- to cut up
Any member of the class Insecta. All organisms in this class are segmented into three body parts, have an exoskeleton, and have three pairs of legs.

Insecticide
Greek
in- in, into, toward, against, on , upon
-secare- to cut up
-cide (caedere) to cut, kill, hack at, or strike

Type of pesticide that controls or eliminates insects that adversely affect plants, animals, or people.

Insectivore
Greek/Latin
in- in, into, toward, against, on, upon
-secare- to cut up
-vorare to eat, devour
Animal or plant that feeds on insects.

Instinct
Latin
instinctus impulse
A complex pattern of innate behavior.

Insulator
Latin
insula- island
-or a conition or property of things or persons
A material that insulates or retards the transfer of energy, especially a nonconductor of sound, heat, or electricity.

Insulin
Latin
insula- island
-in protein or derived from protein
A hormone secreted by the islets of Langerhans in the pancreas. Insulin is essential for the proper uptake and metabolism of sugar.

Integument
Latin
in- in, into, toward, against, on, upon
-tegere to cover
A natural outer covering or coat, such as the skin of an animal or the membrane enclosing an organ.

Interaction
Latin
inter- between, among
-agere- to do
-ion state, process, or quality of
Any of four fundamental ways in which elementary particles and bodies can influence each other, classified as strong, weak, electromagnetic, and gravitational.

Intercellular
Latin
inter- between, among
-cella- chamber
-ar relating to or resembling
Located between cells.

Intercloud gas
Greek/Middle English
inter- between, among
-clud rock, hill
khaos (**Greek**) gas, empty space

Low-density regions of the interstellar medium that fill the space between interstellar clouds.

Intercostal
Latin
inter- between, among
-costo- rib
-al of the kind of, pertaining to, having the form or character of
Situated between the ribs.

Intercrystalline
Latin/Greek
inter- between, among
-krystallinos- rock crystal
-ine of or relating to
Between the crystals of a solid substance.

Interdependent
Latin
inter- between, among
-depend- relying on
-ent causing an action, being in a specific state, within
Mutually dependent; having a direct relationship with one another.

Interferometer
Latin
inter- between, among
-ferir- to strike
*-meter (**metron**)* instrument or means of measuring, to measure
An instrument for measuring very small lengths, distances, and changes in the dimensions, density, and other properties of a substance by means of the interferences of two rays of light.

Interlunar
Latin
inter- among, mutually, together, between
-luna- the moon
-ar relating to or resembling
Pertaining to the period between the old and new moon, during which the moon is not visible from the earth.

Intermolecular
Latin
inter- among, mutually, together, between
-moles- mass
-ule small, tiny
-ar relating to or resembling
Describes forces that are exerted by molecules on each other and that, in general, affect the macroscopic properties of the material of which the molecules are a part.

Internal
Latin
internus- within
-al of the kind of, pertaining to, having the form or character of
Of, relating to, or located within the limits or surface; inner.

Internode
Greek
inter- among, mutually, together, between
-node the point on a plant where a leaf stalk or petiole attaches to the stem
Distance along the stem of a plant between two successive nodes.

Internuclear
Latin
inter- among, mutually, together, between
-nucula- kernel, little nut
-ar relating to or resembling
Located between nuclei.

Interphase
Greek
inter- among, mutually, together, between
-phasis appearance
The stage of cell division during which the chromosomes are uncondensed and are copied.

Interspecific
Greek
inter- among, mutually, together, between
-specif- appearance/kind
-ic (ikos) relating to or having some characteristic of
Refers to a relationship occurring between species.

Interstellar
Latin
inter- among, mutually, together, between
-stella star
Between or among the stars ("interstellar gases").

Interstitial
Latin
inter- among, mutually, together, between
-sistere to stand
Situated in the interstices or spaces between structures such as cells, organs, or grains of sand.

Intertidal zone
Latin/Old English/Greek
inter- (**Latin**) among, mutually, together, between
-tid- (**Old English**) division of time
-alis (**Latin**) of, relating to, characterized by
zone (**Greek**) girdle, celestial zone
The marine zone located in the area of shoreline between high and low tides.

Interval
Latin
inter- among, mutually, together, between
-vallum ramparts
Space between objects.

Intestine
Latin
intestinus within, internal
The tubular portion of the alimentary canal extending from the stomach to the anus; in humans and other mammals, the intestine consists of two segments, the small intestine and the large intestine.

Intracellular
Latin
intra- within, inside
-cellula- chamber
-ar relating to or resembling
Occurring within a body cell or cells.

Intramolecular
Latin
intra- within, inside
-moles- mass
-ule- small, tiny
-ar relating to or resembling
Pertains to the characteristics and properties of any given molecule.

Intraspecific
Latin
intra- within, inside
-specif- appearance/kind
-ic (ikos) relating to or having some characteristic of
Referring to a relationship occurring within a species.

Intrinsic
Latin
intrinsicus- inward
-ic (ikos) relating to or having some characteristic of
Relating to the central or core nature of a thing.

Intron
Latin
intron occurring within a gene
A segment of gene situated between exons that is removed before the translation of messenger RNA.

Introvert
Latin
intr- inwardly, within
-vertere to turn
The anterior narrow portion that can be withdrawn (introverted) into the trunk of a sipunculid worm.

Intrusive
Latin

in- into, on, among
-trudere thrust
Referring to igneous rocks that form at depths below the earth's surface

Invertebrate
Latin
in- without
-vertebratus backbone
Having no vertebrae (backbone).

Inverted
Latin
in- to cause to be
-vertere to turn
Reversed in terms of the position, order, or condition of.

Ionic
Greek
ion- (ienai) to go, something that goes
-ic (ikos) relating to or having some characteristic of
Containing an atom or group of items that have acquired a net electric charge.

Ionization
Greek
ion- (ienai) to go, something that goes
-zation action, process, or quality of
Energy required to remove most loosely held electrons from an atom.

Ionosphere
Greek
ion- (ienai) to go, something that goes
-sphaira a globe shape, ball, sphere
The lower part of the thermosphere, where electrically charged particles called ions are found.

Ipsilateral
Latin
ipse- self, same
-latus- side
-al of the kind of, pertaining to, having the form or character of
Located on or affecting the same side of the body.

Iris
Latin
irid rainbow
In biology, the colored part of the eye that regulates the amount of light allowed into the interior of the eyeball; in botany, the name given to a group of tropical flowering plants; in physics, a diaphragm.

Irrigate
Latin
in- to cause to be
-rigare to water

To supply dry land with water by means of ditches, pipes, or streams; to water artificially.

Isobar
Greek
isos- equal, uniform, same, similar, alike
-baros weight, heavy; atmospheric pressure
Any of the lines on a map joining places that have the same air pressure.

Isobaric
Greek
isos- equal, uniform, same, similar, alike
-baros- weight, heavy; atmospheric pressure
-ic (ikos) relating to or having some characteristic of
Of a thermodynamic process in which a substance experiences no change in pressure.

Isochoric
Greek
isos- equal, uniform, same, similar, alike
-choros- of or having to do with volume
-ic (ikos) relating to or having some characteristic of
Refers to a thermodynamic process in which a substance experiences no change in volume.

Isoelectric
Greek
isos- equal, uniform, same, similar, alike
-elektron- charge, electricity, dealing with positive and negative charges
-ic (ikos) relating to or having some characteristic of
Having an equal number of electrons outside the nucleus.

Isomer
Greek
isos- equal, uniform, same, similar, alike
-meros part, share
Any of two or more nuclei with the same mass number and atomic number that have different radioactive properties and can exist in any of several energy states for a measurable period of time.

Isometric
Greek
isos- equal, uniform, same, similar, alike
-metr- measurement
-ic (ikos) relating to or having some characteristic of
Equal in dimension or measurement; in biology, relating to the contraction of muscles against an immovable resistant force, where the length of the muscle fibers remains the same.

Isopod
Greek
isos- equal, uniform, same, similar, alike
-pod foot

Any of numerous crustaceans of the order Iso-poda, characterized by a flattened body bearing seven pairs of legs, and including the sow bugs and gribbles.

Isotactic

Greek

isos- equal, uniform, same, similar, alike
-taktos ordered

Describes the orientation of the methyl groups on a polypropylene chain in plastics, which in this case is all on the same side.

Isotherm

Greek

isos- equal, uniform, same, similar, alike
-thermos- combining form of "hot" (heat)

In meteorology, a line drawn on a weather map indicating points of equal temperature.

Isotonic

Greek

isos- equal, uniform, same, similar, alike
-ton- tension
-ic (ikos) relating to or having some characteristic of

Of equal tension; having the same concentration of solute on both sides of a membrane.

Isotope

Greek

isos- equal, uniform, same, similar, alike
-topos place

One of two or more atoms having the same atomic number but different mass numbers.

Isthmus

Greek

isthmos narrow neck

In biology, a narrow strip of tissue connecting two parts or lobes of a gland or organ; in earth science, a narrow strip of land connecting two larger sections of land.

J

Jaundice
> Latin
> *galbinus* yellowish
> Yellow discoloration of the eyes, mucous membranes, and skin caused by deposits of bile, usually as a result of a disease, such as hepatitis.

Jejunum
> Latin
> *ieiunus* fasting (referring to its always being found empty when dissected)
> The very large section of small intestine beginning at the end of the duodenum and ending at the beginning of the ileum.

Joule
> Old English
> *Joule* English physicist (James Prescott Joule) who developed the first law of thermodynamics
> A unit of electrical energy equal to 10 million ergs or one newton-meter.

Jurassic
> French
> *jurassique/jura-* mountains
> *-ic (ikos)* relating to or having some characteristic of
> Of or belonging to the geologic time, rock series, or sedimentary deposits of the second period of the Mesozoic era, in which dinosaurs continued to be the dominant land fauna and the earliest birds appeared.

Juvenile
> Latin
> *iuvenis-* young
> *-ile* changing
> Not fully grown or developed; young.

K

Kalemia
Latin
kalium- potassium
-haima- blood
-ia names of diseases, place names, or Latinizing plurals
The presence of excessive amounts of potassium in the blood.

Kame
Middle English
camb comb
A short ridge or mound of sand and gravel deposited during the melting of glacial ice.

Karyapsis
Greek
kary- nut, walnut, kernel, nucleus
-haptien to fasten, join
The process of the fussion or union of nuclei in conjugating cells.

Karyochrome
Greek
kary- nut, walnut, kernel, nucleus
-chrome pigment
A nerve cell whose nucleus is deeply stainable although its body is not.

Karyocyte
Greek
kary- nut, walnut, kernel, nucleus
-cyte (kutos) sac or bladder that contains fluid
The term for any cell possessing a nucleus.

Karyogamic
Greek

kary- nut, walnut, kernel, nucleus
-gam- husband or wife; to marry
-ic (ikos) relating to or having some characteristic of
Describes a process pertaining to or characterized by the union of two nuclei.

Karyogamy
Greek
kary- nut, walnut, kernel, nucleus
-gam- husband or wife; to marry
-y place for an activity, condition, state
The fusion of two cell nuclei following plasmogamy during fertilization.

Karyogenesis
Greek
kary- nut, walnut, kernel, nucleus
-gen- to give birth, kind, produce
-sis action, process, state, condition
The growth and development of the nucleus of a cell.

Karyokinesis
Greek
kary- nut, walnut, kernel, nucleus
-kinetikos- to move; set in motion
-sis action, process, state, condition
A phenomenon involved in the division of the nucleus, usually an early stage in the process of cell division, or mitosis.

Karyoklasis
Greek
kary- nut, walnut, kernel, nucleus
-klastos- break, break in pieces
-sis action, process, state, condition

The breaking down of the cell nucleus or nuclear membrane.

Karyolymph
Greek
kary- nut, walnut, kernel, nucleus
-lympha clear water, water nymph
The liquid part of a cell nucleus, as contrasted with the chromatin and linin.

Karyolysis
Greek
kary- nut, walnut, kernel, nucleus
-ly- (luein) to loosen, dissolve, dissolution, break
-sis action, process, state, condition
Form of necrobiosis in which the nucleus of a cell swells and gradually loses its chromatin.

Karyomegaly
Greek
kary- nut, walnut, kernel, nucleus
-megas- large, great, big, powerful
-ly like, likeness, resemblance
Abnormal enlargement of the nucleus of a cell, not caused by polyploidy.

Karyometry
Greek
kary- nut, walnut, kernel, nucleus
-metria (metron) the process of measuring
The measurement of a cell nucleus.

Karyomorphism
Greek
kary- nut, walnut, kernel, nucleus
-morph- shape, form, figure, or appearance
-ism state or condition, quality
The shape of a cell nucleus.

Karyophage
Greek
kary- nut, walnut, kernel, nucleus
-phagos (phagein) to eat, eating
A protozoan that is capable of phagocytic action on the nucleus of the cell it infects.

Karyoplasm
Greek
kary- nut, walnut, kernel, nucleus
-plasm (plassein) to mold or form cells or tissues
The nucleoplasm or protoplasm of the nucleus of a cell.

Karyoreticulum
Greek
kary- nut, walnut, kernel, nucleus
-reticul- net or networklike
-um (**singular**) structure
-a (**plural**) structure

The fibrillar part of the karyoplasm as distinguished from the fluid part of karyolymph.

Karyorrhexis
Greek
kary- nut, walnut, kernel, nucleus
-rhxis action or process of bursting
Rupture of the cell nucleus in which the chromatin disintegrates into formless granules that are extruded from the cell.

Karyotype
Greek
kary- nut, walnut, kernel, nucleus
-typos impression, figure
Representation of individual chromosomes cut out from a photograph and grouped together.

Karyozoic
Greek
kary- nut, walnut, kernel, nucleus
-zoon- animal, animal like
-ic (ikos) relating to or having some characteristic of
Existing in or inhabiting the nuclei of cells, as certain protozoa.

Katolysis
Greek
kato- below
-ly- (luein) to loosen, dissolve; dissolution, break
-sis action, process, state, condition
The incomplete or intermediate conversion of complex chemical bodies into simpler compounds; applied especially to digestive processes.

Keel
Old Norse
kjolr ship
Anything with a shape or purpose similar to that of a ship's keel in supporting the whole frame, as in the breastbone of birds.

Keratin
Greek
keras- horn
-in protein or derived from protein
A scleroprotein found in epidermal tissues and modified into hard structures such as horns, hair, and nails.

Ketone
German (from Latin)
keton short for *aketon* or *acetone* (*acetone* is derived from Latin *acetum* [vinegar])
Any of a class or organic compounds having a carbonyl group linked to a carbon atom in each of two hydrocarbon radicals.

Kilogram

Greek
khilioi- thousand
-gramma small weight
A metric unit for the measurement of mass.

Kiloliter

Greek
khilioi- thousand
-litra unit of weight or capacity
A metric unit for the measurement of weight or capacity; usually associated with liquids.

Kilometer

Greek
khilioi- thousand
-meter (metron) instrument or means of measuring; to measure
A metric unit for the measurement of distance.

Kindling

Old Norse
kynda- cause or to give birth to
-ing the act of or action
Substances such as wook chips, dried sticks, or charcoal that are relatively easy to ignite.

Kinematics

Greek
kinemat- mechanics of movement
-ic (ikos) relating to or having some characteristic of
The branch of mechanics that studies the motion of a body, or a system of bodies, with no consideration given to the body's mass or the forces acting on it.

Kinetic

Greek
kinetikos- to move; set in motion
-ic (ikos) relating to or having some characteristic of
The kind of energy relating to or produced by motion.

Kinetochore

Greek
kinetos- moving
-khoros place
Structure that forms on the centromere during mitosis for binding microtubules.

Kinetosome

Greek
kinetikos- to move; set in motion
-soma (somatiko) body
The self-duplicating granule at the base of the flagellum or cilium; similar to the centriole; also called basal body or blepharoplast.

Kingdom

Old English
cyning- principal, chief
-dom property, jurisdiction
In biology, the highest level in the hierarchy of the taxonomical classification of living organisms.

Kyphosis

Latin
kuphos- humpbacked, bent over
-sis action, process, state, condition
Exaggerated thoracic curvature.

L

Label
Middle English
lap- to wrap, to fold
-elle diminutive
To infuse or treat a substance with a radioactive isotope or a fluorescent dye so that its course of activity can be traced through a series of reactions; usually done in a living organism.

Labrum
Latin
labr- lip
-um (**singular**) structure
-a (**plural**) structure
A structure forming the roof of the mouth in insects.

Labyrinthodont
Greek
labyrinthos- labyrinth, inner ear, double-headed axe, of Lydian origin
-odontos tooth
A group of Paleozoic amphibians containing the temnospondyls and the anthracosaurs.

Labyrinthus
Greek
labyrinthos- labyrinth, inner ear, double-headed axe, of Lydian origin
-us thing
The portion of the inner ear characterized by the semicircular canals and involved with hearing and balance.

Laccolith
Greek
lakkos- cistern
-lith rock, stone

A mass of igneous rock intruded between layers of sedimentary rock, resulting in uplift.

Lactescence
Latin
lac- milk or lactic acid
-escence giving off light of the kind or type specified
A milky appearance; milkiness

Lactic
Latin
lac- milk or lactic acid
-ic (ikos) relating to or having some characteristic of
Of or pertaining to milk; procured from sour milk or whey, as in lactic acid; lactic fermentation.

Lactose
Latin/Greek
lac- milk or lactic acid
-ose sugar, carbohydrate
A disaccharide found in the milk of all mammals; a sugar found in milk that breaks down into glucose and galactose, and creates lactic acid through fermentation.

Lacuna
Latin
lacuna lagoon
A space or cavity in bone that is occupied by a bone cell or a cartilage cell.

Lagomorph
Greek
lagos- hare
-morph shape, form, figure, or appearance
Gnawing, herbivorous mammals, including rabbits, hares, and pikas.

Lake
Latin
lacus lake
A large inland body of freshwater or salt water.

Lamella
Latin
lamin- thin plate or layer, neurophysis of a vertebra
-ella dimunitive
A thin layer of bony matrix material.

Laminectomy
Latin/Greek
lamin- thin plate or layer, neurophysis of a vertebra
-ekt- outside, external, beyond
-tomos (temnein) to cut, incise, section
Surgical removal of the posterior arch of a vertebra.

Laparonephrectomy
Greek
lapar- the soft part of the body between the ribs, hip, and flank; the loin
-nephr- kidney
-ekt- outside, external, beyond
-tomos (temnein) to cut, incise, section
Removal of the kidney by an incision in the loin.

Laparosalpingo-oophorectomy
Greek
lapar- the soft part of the body between the ribs, hip, and flank; the loin
-salping- tube, trumpet
-oophor- ovary
-ekt- outside, external, beyond
-tomos (temnein) to cut, incise, section
Removal of the Fallopian tube and ovary through an abdominal incision.

Laparotomy
Greek
lapar- the soft part of the body between the ribs, hip, and flank; the loin
-tomos (temnein) to cut, incise, section
The act of cutting through the abdominal wall into the cavity of the abdomen.

Larvae
Latin
larva mask, specter
The intermediary stage of development in insects and many other animals between the egg and adult stages. Referred to as a larva because the adult stage is hidden or masked.

Laryngitis
Greek
larunx- part of the respiratory system in the neck, cartilage, muscular tube
-itis inflammation, burning sensation
Inflammation of the larynx, often with a temporary loss of voice.

Lateral
Latin
lateralis side
Of, relating to, or being situated at or on the side.

Latitude
Latin
latus- wide
-tudo condition, state, quality
The angular distance north or south of the earth's equator, measured in degrees along a meridian, as on a map or globe.

Lattice
Germanic
latte lathe
A regular, periodic configuration of points, particles, or objects throughout an area or a space, especially the arrangement of ions or molecules in a crystalline solid.

Lava
Latin
labi to fall
Molten rock that reaches the surface of the earth through a fissure of a volcano.

Leach
Late Middle English
leche to wet or to infuse
To dissolve out soluble parts from, by running water or other liquid through slowly.

Leaf
Old English
leaf leaf
Typically green, flattened structure of a plant that is attached to a stem. It serves as the primary structure for energy production via photosynthesis.

League
Latin
leuga a measure of distance
A unit of distance equal to 3.0 statute miles (4.8 kilometers).

Lepidoptera
Greek
lepidos- scale, flake
-ptera feather, wing
The order of insects that includes butterflies and moths.

Lepidosaurs
Latin
lepidos- scale, flake
-sauros lizard

A lineage of diapsid reptiles that appeared in the Permian period and includes the modern snakes, lizards, amphisbaenids, and tuataras, as well as the extinct ichthyosaurs.

Leprosy
Latin
lepra- flake, scale, scaly, scabby
-y place for an activity, condition, state
A slowly progressive, chronic infectious disease characterized by granulomatous or neurotrophic lesions in the skin, mucous membranes, nerves, bones, and viscera, with a broad spectrum of clinical symptoms.

Leptocephalus
Greek
leptos- thin
-kephale- head
-us thing
Transparent, ribbonlike migratory larva of the European or American eel.

Lepton
Greek
leptos- small or fine
-on a particle
Any of a family of elementary particles that participate in a weak interaction, including the electron, the muon, and their associated neutrinos.

Lethal
Latin
letum death
Relating to or capable of causing death.

Leuco
Greek
leukos white, clear, or colorless
Of or designating a reduced, colorless form of a dye that is fixed on a fiber and then reconstituted into the dye by means of oxidizing agents.

Leucoplast
Greek
leukos- white, clear, or colorless
-plastos (plassein) something molded; to mold
A colorless plastid in the cytoplasm of plant cells around which starch collects.

Leukemia
Greek
leukos- white, clear, or colorless
-haima- blood
-ia names of diseases, place names, or Latinizing plurals
A form of cancer characterized by uncontrolled production of abnormal white blood cells.

Leukoblast
Greek
leukos- white, clear, or colorless
-blastos bud, germ cell
An immature white blood cell; also called a proleukocyte.

Leukocyte
Greek
leukos- white, clear, or colorless
-kutos (cyto) sac or bladder that contains fluid
White blood cell, of which there are several types, each having a specific function in protecting the body from invasion by foreign substances and organisms.

Leukocytopenia
Greek
leukos- white, clear, or colorless
-kutos- (cyto) sac or bladder that contains fluid
-penia reduction, poverty, lack, deficiency
A condition in which there is a decrease in or an insufficiency of white blood cells circulating in the body.

Leukocytosis
Greek
leukos- white, clear, or colorless
-kutos- (cyto) sac or bladder that contains fluid
-osis action, process, state, condition
An increase in the number of white blood cells in the circulating blood.

Leukopenia
Greek
leukos- white, clear, or colorless
-penia reduction, poverty, lack, deficiency
A condition in which the number of white blood cells circulating in the blood is abnormally low.

Leukosarcoma
Greek
leukos- white, clear, or colorless
-sarko- flesh, meat
-oma tumor, neoplasm
A type of lymphoma characterized by large numbers of abnormal lymphocyte precursors in the blood.

Levator
Latin
levare- to lift, raise
-or a condition or property of things or persons; person who does something
Any muscle that elevates a part of the body.

Lever
Latin
levis light
A simple machine consisting of a rigid bar pivoting on a fixed point and used to transmit force, as

in raising or moving a weight at one end of a beam by pushing down on the other end.

Levorotatory
Latin
laevus- left or counterclockwise
-rota- wheel
-ory of or pertaining to
Rotating to the left in a plane of polarized light.

Libration
Latin
libra- balance
-ion state, process, or quality of
A very slow oscillation, real or apparent, of a satellite as viewed from the larger celestial body around which it rotates.

Lichen
Greek
leikhein to lick
A plantlike organism consisting of a symbiotic relationship between algae and fungi; usually found on rocks and other regions with minimal sources of food or water.

Life
Old English
lif life
The term designating any physiologically active organism; the capacity to carry on all life processes.

Ligament
Latin
ligare- to bind, tie
-ment causing an action, or being in a specific state
A strong, elastic connective tissue that crosses a joint and prevents excessive movement that could dislocate the joint.

Ligant
Latin
ligare to bind, tie
A charged or uncharged molecule that can bind to a metal molecule or ion and form a large, complex ion.

Ligroin
German
ligroin ligroin
Petroleum ether; a volatile, flammable liquid mixture of hydrocarbons obtained by the fractional distillation of petroleum; used as a solvent.

Limicole
Latin
limus- mud, slime
-cole inhabit
Living in mud; a group of shore bird such as the sandpipers or plovers.

Limivorous
Latin
limus- mud, slime
-vorare eat, swallow
Feeding on mud for the organic matter it contains; characteristic of certain amnelids,

Limnetic
Greek
limne- lake
-ic (ikos) relating to or having some characteristic of
Relating to of having the characteristic of living in the deep waters of a lake or pond.

Lingual
Latin
lingua- tongue, language
-al of the kind of, pertaining to, having the form or character of
Of or pertaining to the tongue or tonguelike organ.

Lipid
Greek/French
lipos- fat
-ide group of related chemical compounds
Any group of organic compounds, including fats, oils, waxes, sterols, and glycerides, that are insoluble in water but soluble in organic solvents.

Liposome
Greek
lipos- fat
-soma (somatiko) body
Droplet of phospholipid molecules formed in a liquid environment.

Liquefy
Latin
liquere- flow, fluid, wave; to be liquid
-fy (ficare) make, do, build, produce
To cause to become liquid, especially to melt (a solid) by heating or to condense (a gas) by cooling.

Liquid
Latin
liquere- flow, fluid, wave; to be liquid
-id state, condition; having, being, pertaining to, tending to, inclined to
Matter that has a distinct volume but no specific shape.

Lithium
Greek
lithos- stone, rock
-ium quality or relationship
A silvery-colored soft metal with the atomic number 3. It is used as a therapeutic for bipolar, depressive disorders. It is also used as a heat transfer medium and is found in various alloys, ceramics, and glass.

Lithosphere
Greek
lithos- stone or rock
-sphaira a globe shape, ball, sphere
The solid outer layer of the earth, consisting of the crust and upper mantle.

Lithotomy
Greek
lithos- stone or rock
-tomos (temnein) to cut, incise, section
The surgical removal of a stone from the urinary tract.

Lithotripsy
Greek
lithos- stone or rock
-tripsy (tribein) to crush; massage, rub, rubbing, friction, grind
Surgical crushing of stones, as in the bladder or ureters.

Litmus
Middle Dutch
leken- to drip
-mosi moss
A blue coloring matter obtained from lichens, used as an acid/base indicator. It turns red in an acidic pH of 4.5 and turns blue in bases at pH 8.3.

Littoral
Latin
litoralis pertaining to the seashore
On the shore, coastal; a zone between high and low tides.

Lobopodium
Greek
lobos- rounded projection, especially a rounded projecting anatomical part
-podos- foot
-ium quality or relationship
Blunt, lobelike pseudopodium.

Lobotomy
French/Greek
lobos- rounded projection, especially a rounded projecting anatomical part
-tomos (temnein) to cut, incise, section
Surgical incision into the frontal lobe of the brain to sever one or more nerve tracts. This technique was formerly used to treat certain mental disorders but now is rarely performed.

Locomotion
Latin
locus- a place or location
movere- to move
-ion state, process, or quality of

The ability of an organism to move from one place to another place.

Lodestone
Old English
lad- way
-stan stone, rock
Magnetite, a common ore that is a natural magnet. At one time it was used by sailors to navigate.

Loess
German
losch loose
A buff to gray windblown deposit of fine-grained calcareous silt or clay.

Longitude
Latin
longus- long
-tude state or quality
Angular distance on the earth's surface, measured east or west from the prime meridian at Greenwich, England, to the meridian passing through a particular position; expressed in degrees (or hours), minutes, and seconds.

Lophophile
Greek
lophos- crest
-phile one who loves or has a strong affinity or preference for
Thriving on hilltops; hilltop plants, plant communities existing on hilltops.

Lophophore
Greek
lophos- crest
-phoros bearing
Tentacle-bearing ridge or arm within which is an extension of the coelomic cavity in lophophorate animals (ectoprocts, brachiopods, and phoronids).

Lophophyte
Greek
lophos- crest
-phyte plant
Plants that thrive on hilltop or crest environments.

Lophotrichous
Greek
lopho- ridge, crest
-tricho- hair
-ous full of, having the quality of, relating to
Refers to having two or more flagella at one end of a cell.

Lordosis
Latin
lordos- to bend backward
-sis action, process, state, condition

An abnormal, exaggerated curvature of the vertebral column in the lumbar region.

Lumbar
Latin
lumbus loin
Relating to the lower back or small of the back.

Lumen
Latin
lumen an opening, light
In biology, the space or cavity within an organ or organ system, such as within blood vessels or the alimentary canal. In physics, the amount of light given out through a solid angle by a source of one candela intensity, radiating equally in all directions.

Luminous
Latin
lumen- an opening, light
-ous full of, having the quality of, relating to
Describes an object or living thing that has the capacity to emit light, or glow.

Lunar
Latin
luna- the moon
-ar relating to or resembling
Of, involving, caused by, or affecting the moon.

Lunarscape
Latin
luna- the moon
-scapus scene, view
Landscape of rock similar to the surface of the moon.

Lunation
Latin
luna- the moon
-ation act or process
The period between new moons: 29 days, 12 hours, and 44 minutes.

Luster
Latin
lustrare light, illuminate
Shining or being reflected by light.

Lymph
Latin
lympha clear water, water nymph
Fluid, derived from tissue fluid, that is carried in lymphatic vessels.

Lymphatic
Greek
lympha- clear water, water nymph
-ic (ikos) relating to or having some characteristic of
Of or relating to lymph, a lymph vessel, or a lymph node.

Lymphocyte
Greek/Latin
lympha- clear water, water nymph
-cyte (kutos) sac or bladder that contains fluid
Specialized white blood cell that occurs in two forms: T lymphocyte and B lymphocyte.

Lymphoma
Greek
lympha- clear water, water nymph
-oma tumor
Any of various usually malignant tumors that arise in the lymph nodes or in other lymphoid tissue.

Lysogenic
Greek
ly- (luein) to loosen, dissolve, dissolution, break
-gen- to give birth, kind, produce
-ic (ikos) relating to or having some characteristic of
Capable of causing or undergoing lysis.

Lysosome
Greek
ly- (luein) to loosen, dissolve, dissolution, break
-soma (somatiko) body
A cytoplasmic, membrane-bound particle containing hydrolytic enzymes that function in intracellular digestive processes.

Lysozyme
Greek
ly- (luein) to loosen, dissolve, dissolution, break
-zume fermenting, leaven
An enzyme occurring naturally in egg white, human tears, saliva, and other body fluids and capable of destroying the cell walls of certain bacteria and thereby acting as a mild antiseptic.

M

Macradenous
Greek
makros- long, large, great
-aden- lymph gland(s)
-ous full of, having the quality of, relating to
Having large glands.

Macrencephaly
Greek
makros- long, large, great
-enkephalos- in the head
-ly like, likeness, resemblance
Overgrowth of the brain.

Macrocardius
Greek
makros- long, large, great
-kard- heart, pertaining to the heart
-us thing
A fetus with an extremely large heart.

Macroevolution
Latin
makros- long, large, great
-evolvere to unfold
Evolutionary change on a grand scale, encompassing the origin of novel designs, evolutionary trends, adaptive radiation, and mass extinction.

Macrogamete
Greek
makros- long, large, great
-gamos marriage
The larger of the two gamete types in a heterogametic organism, considered the female gamete.

Macroglobulin
Greek
makros- long, large, great
-globu- globe
-in of or derived from a protein
An immunoglobulin of very high molecular weight, usually above 900,000.

Macronucleus
Greek
makros- long, large, great
-nucula- kernel, little nut
-us thing
Large nucleus that controls the functions of the cell.

Macrophage
Greek
makros- long, large, great
-phagos (phagein) to eat, eating
A large white blood cell that can engulf hundreds of bacteria.

Macrovolt
Greek
makros- long, large, great
-volt electric potential
Large electric potential (one million volts).

Madreporite
Latin
madre- mother
-pora- passageway
-ite component of a part of a body
A perforated. platelike structure in most echinoderms that forms the intake for their water-vascular systems.

Mafic

Latin

ma- the element magnesium

-ic (ikos) relating to or having some characteristic of

Containing or relating to a group of dark-colored minerals that are composed chiefly of magnesium and iron in igneous rock.

Magma

Greek

mag- to knead

-ma form or character of

The name given to molten rock under the surface of the earth. Magma becomes lava if it escapes from a volcano to the earth's surface.

Magnet

Greek

magnes stone from Magnesia (city in Asia Minor)

An object that is surrounded by a magnetic field and that has the property, either natural or induced, of attracting iron or steel.

Magnetosphere

Greek

magnes- stone from Magnesia (city in Asia Minor)

-sphaira a globe shape, ball, sphere

Region around an object where the influence of the object's magnetic field can be felt.

Magnification

Latin/Greek

magn- great

-fic- to make

-ion state, process, or quality of

The process of making things look larger.

Magnitude

Latin

magnu- large

-tude state, quality, condition of

The overall size of a quantity.

Malacoderm

Greek

malacia- softening of tissue

-derm skin

Having soft skin or soft flexible bodies, as is characteristic of fireflies.

Malacopterygia

Greek

malacia- softening of tissue

-pterug- wing

-ia names of diseases, place names, or Latinizing plurals

Order of fishes where the fins are soft and closely jointed; carp is an example.

Malacosarcosis

Greek

malacia- softening of tissue

-sarko- flesh, meat

-sis action, process, state, condition

Softness of muscular tissue.

Malacostracan

Greek

malako- soft

-ostracon shell

Any member of the crustacean subclass Malacostraca, which includes both aquatic and terrestrial forms of crabs, lobsters, shrimps, pillbugs, sand fleas, and others.

Malaria

Italian

mala- bad

-aria air

Air infected with a noxious substance capable of causing disease.

Malignant

Latin

malignus bad, attach, malign

Relates to a disease that is threatening to life; virulent; cancerous.

Malleable

Latin

malleus- hammer

-able capable, be inclined to, tending to, given to

A property of metal enabling it to be pounded or rolled into thin sheets.

Mallophaga

Greek

mallos- wool

-phagos (phagein) to eat, eating

Chewing lice; extensive group of small insects that are parasitic in nature on birds and mammals and feed on feathers and hair.

Malnutrition

Latin

mala- bad

-nutrire- to suckle, nourish

-ent causing an action, being in a specific state, within

Poor nutrition related to or caused by an insufficient or poorly balanced diet, faulty digestion, or faulty use of foods.

Maltase

Greek

malt- seed or grain

-ase indicating an enzyme

Enzyme in plants and animals that breaks down disaccharide maltose into glucose.

Maltose
Greek
malt- seed or grain
-ose sugar, carbohydrate
Disaccharide sugar in which both monosaccharide parts are glucose.

Mammal
Latin
mamma- breast
-al of the kind of, pertaining to, having the form or character of
An animal with hair that feeds its young with milk from mammary glands.

Mammary
Greek
mamma- breast
-ary of, relating to, or connected with
Of or relating to the breasts (e.g., mammary glands).

Mandible
Latin
mandere to chew
The lower jaw of vertebrates.

Mantle
Latin
mantellum layer
In geology, the layer of earth between the central molten core and the surface crust.

Manubrium
Latin
manus- hand
-ium quality or relationship
A bony segment of the sternum shaped like a handle.

Marine
Latin
mare sea
Of or relating to the sea.

Marsupial
Greek
marsuppos- pouch or purse
-ial (variation of *-ia*) relating to or characterized by
Mammal that bears its immature young in a marsupium, or pouch.

Mass
Greek
maza mass, large, amount
The property of a body that is a measure of its inertia; commonly taken as a measure of the amount of material the body contains and that causes it to have weight in a gravitational field.

Mastication
Greek
mastikhan- to grind the teeth
-ion state, process, or quality of
The process of using one's teeth to chew and grind food.

Mastoid (process)
Greek
mastos- breast
-oid (oeides) resembling; having the appearance of
A small process resembling a nipple that is found on the temporal bone.

Matter
Latin
materia substance from which something is made
Something that occupies space and can be perceived by the senses; a physical substance or the physical universe as a whole.

Maxilla
Latin
maxilla jawbone
The fusion of two bones in mammals forming the upper jaw.

Maxilliped
Latin
maxilla- jawbone
-ped foot
One of the pairs of head appendages located just posterior to the maxilla in crustaceans; a thoracic appendage that has become incorporated into the feeding mouthparts.

Maxima
Latin
maximus greatest
The greatest values assumed by a function over a given interval.

Mean
Old English
maenan to tell of
The average of a group of sample numbers as calculated by dividing the sum of the numbers by the number of samples.

Meatus
Latin
meare to pass
An opening or a canal—for example, the external auditory meatus.

Mechanical
Greek
mekhane- machine, device
-al of the kind of, pertaining to, having the form or character of
Relating to a machine or the functionality of a machine. Mechanical advantage refers to the measurement of the output force of the machine (lever) versus the input force.

Meconium
Greek
mekonion poppy juice
The first feces of the newborn; the coloration is usually greenish black to light brown.

Median
Latin
medius middle
The average that gives the midpoint of a range or distribution.

Medium
Latin
medius middle
An intervening substance through which something else is transmitted or carried.

Medulla
Latin
merulla middle
The inner core of certain structures or organs.

Medusa
Latin
medein to protect
Tentacled, bell-shaped, free-swimming body plan of cnidarians.

Megalocephaly
Greek
megal- large, great
-kephalikos head
A birth defect that causes an abnormally large head.

Megaspore
Greek
megas- large, great, big, powerful
-spora seed
In plants, a haploid (n) spore that develops into a female gametophyte.

Meiosis
Greek
meion- smaller, less
-sis action, process, state, condition
The cellular process that results in the number of chromosomes in gamete-producing cells being reduced to one-half, and that involves a reduction division, in which one of each pair of homologous chromosomes passes to each daughter cell, and a mitotic division.

Melanin
Greek
melas- the color black, dark
-in protein or derived from protein
Dark brown pigment of many animals, giving brown and yellow coloration to skin and/or hair.

Melanocyte
Greek
melas- the color black, dark
-cyte (kutos) sac or bladder that contains fluid
An epidermal cell capable of synthesizing melanin.

Melanoderma
Greek
melas- the color black, dark
-derma skin
Black or dark skin coloring (pigmentation); literally, black skin.

Melanoma
Greek
melas- the color black, dark
-oma community
A dark-pigmented, usually malignant tumor arising from a melanocyte and occurring most commonly in the skin.

Membrane
Latin
membrana thin skin
Thin layer of tissue composed of epithelial cells and connective tissue that covers a surface.

Meningitis
Greek
mening- meninx
-itis inflammation, burning sensation
Inflammation of the meninges of the brain and the spinal cord, most often caused by a bacterial or viral infection.

Meniscus
Greek
mensikos moon, month
The concave or convex upper surface of a nonturbulent liquid in a container.

Meridian
Latin
medius- middle
-die day
In astronomy, a great circle passing through the two poles of the celestial sphere and the zenith of a given observer.

Meristem
Greek
meristos- divided
-en to make or cause
The undifferentiated plant tissue from which new cells are formed, as that at the tip of a stem or root.

Mesentery
Greek
mesos- middle
-enteron gut
A membrane that suspends many of the organs of vertebrates inside fluid-filled body cavities.

Mesoderm
Greek
mesos- middle
-derma skin
The germ layer formed between the ectoderm and the endoderm of an embryo.

Mesoglea
Greek
mesos- middle
-gloia glue
The clear, inert, jellylike substance that makes up the majority of the bodies of jellyfish, comb jellies, and certain other primitive sea creatures.

Mesomorphic
Greek
mesos- middle
-morph- shape, form, figure, or appearance
-ic (ikos) relating to or having some characteristic of
Existing in a state of matter intermediate between liquid and crystal; describes any individual having the characteristics of a stout, healthy physique developed from the embryonic mesomorphic layer.

Meson
Greek
mesos- middle
-on a particle
The class of elementary particles with masses between baryons and leptons.

Mesophyll
Greek
mesos- middle
-phullon leaf
The ground tissue of a leaf, sandwiched between the upper and lower epidermis and specialized for photosynthesis.

Mesophyte
Greek
mesos- middle
-phyte plant
A plant that has adapted to grow in areas having moderate moisture conditions.

Mesosphere
Greek
mesos- middle
-sphaira a globe shape, ball, sphere
The zone of the earth's interior that extends from the lithosphere to the core.

Mesozoic
Greek
mesos- middle
-zoikos- of animals
-ic (ikos) relating to or having some characteristic of
An era of geologic time between the Paleozoic and the Cenozoic, occurring between 248 and 65 million years ago.

Metabolism
Greek
meta- between, after, beyond, later
-bol- (ballein) to put or throw
-ism state or condition, quality
The complex of physical and chemical processes involved in the maintenance of life.

Metacarpus
Greek
meta- between, after, beyond, later
-karpos- wrist
-us thing
The part of the human hand that includes the five bones between the fingers and the wrist.

Metagalaxy
Greek
meta- between, after, beyond, later
-galakt milk
The assemblage of all the galaxies.

Metal
Greek
metallon- mine, ore, quarry, any of a category of electropositive elements from metallum
Any member of the class of substances represented by gold, silver, copper, iron, and tin.

Metallic
Latin/Greek
metallon- mine, ore, quarry, any of a category of electropositive elements from metallum
-ic (ikos) relating to or having some characteristic of
Having characteristics of metals.

Metalloid
Latin/Greek
metallon- mine, ore, quarry, any of a category of electropositive elements from metallum
-oid (oeides) resembling; having the appearance of

A nonmetallic element, such as arsenic, that has some of the chemical properties of a metal.

Metallurgy
Latin/Greek
metallon- mine, ore, quarry, any of a category of electropositive elements from metallum
-ourgos worker
The science and technology involving the study of metals.

Metamere
Greek
meta- between, after, beyond, later
-meros part
Condition of being made up of serially repeated parts; serial segmentation.

Metamorphic
Latin/Greek
meta- between, after, beyond, later
-morph- shape, form, figure, or appearance
-ic (ikos) relating to or having some characteristic of
Refers to a change of physical form, structure, or substance, especially rock that has changed from its original form through the application of heat and pressure.

Metamorphosis
Greek
meta- between, after, beyond, later
-morph- shape, form, figure, or appearance
-osis action, process, state, condition
A change in the form of an animal during normal development after the embryonic stage.

Metaphase
Greek
meta- between, after, beyond, later
-phaseis appearance
The stage of mitosis and meiosis where chromosomes align along the metaphase plate.

Metapopulation
Greek/Latin
meta- between, after, beyond, later
-populus- the people
-ion state, process, or quality of
A population subdivided into several small and isolated populations as a result of habitat fragmentation.

Metatarsus
Greek
meta- between, after, beyond, later
-tarsos- instep
-us thing
The middle part of the human foot that forms the instep and includes the five bones between the toes and the ankle.

Metatheria
Greek
meta- between, after, beyond, later
-ther- wild animal
-ia names of diseases, place names, or Latinizing plurals
Infraclass of marsupial mammals.

Metathesis
Greek
meta- between, after, beyond, later
-tithenai to transpose, to place
A chemical reaction in which a double decomposition occurs, causing parts of two reacting structures to swap places.

Meteor
Greek
meteoron things in air
The luminous phenomenon observed when a meteor enters the atmosphere.

Meteorite
Greek
meteoron- things in air
-ite minerals and fossils
A metallic or mineral mass that has fallen to earth from space.

Meteorologist
Latin/Greek
meteoron- things in air
-ologist one who deals with a specific topic
A person who is a specialist in the study of the weather, the atmosphere, and forecasting.

Meteorology
Latin/Greek
meteoron- things in air
-logy (logos) used in the names of sciences or bodies of knowledge
The study of earth's atmosphere, weather, and climate.

Meter
Greek
meter (metron) instrument or means of measuring; to measure
A metric unit used in the measurement of length equivalent to 39.37 inches.

Methanogens
Greek
methano- methane
-gen to give birth, kind, produce
Organisms that require anaerobic conditions and that produce methane gas.

Methionine
Greek
meth- containing a methyl group
-thio- compound containing sulfur
-ine in a chemical substance
A sulfur-containing amino acid.

Micaceous
Latin
mica- grain
-ous full of, having the quality of, relating to
Pertaining to or containing mica; a laminar rock structure much like mica.

Micelle
Latin
mica- grain, crumb
-elle diminutive
A unit in colloids composed of complex molecules that can alter size without chemical change.

Microbiologist
Greek
mikros- small
-bios- life, living organisms, or tissue
-ologist one who deals with a specific topic
One who specializes in the science of microbiology.

Microbiophagy
Greek
mikros- small
-bios- life, living organisms, or tissue
-phagia eat, eating; consume, ingest
Destruction or lysis of microorganisms by a phage.

Microcephalic
Greek
mikros- small
-cephalo- (kephalikos) head
-ic (ikos) relating to or having some characteristic of
Having a small head or a small cranial cavity.

Microfilaments
Greek/Latin
mikros- small
-filum- thread
-ent causing an action, being in a specific state, within
Any of the minute fibers throughout the cytoplasm of a cell that function primarily in maintaining its structural integrity.

Microfilaria
Greek
mikros- small
-filum- thread
-ia names of diseases, place names, or Latinizing plurals
The minute larval form of the slender, threadlike filarial worm.

Micrometer
Greek
micro- denotes one-millionth of a part
-meter (metron) instrument or means of measuring; to measure
One-millionth of a meter, symbol μm; used in many types of microscopic science, such as cellular biology.

Microneme
Greek
mikros- small
-nema thread
One of the types of structures composing the apical complex in the phylum Apicomplexa; these structure are slender and elongate, leading to the anterior, and thought to function in host cell penetration.

Microorganism
Greek
mikros- small
-organ- complex structure; tool
-ism state or condition, quality
A very small living thing.

Microprocessor
Greek/Latin
mikros- small
-processus- setting out, series of steps
-or a condition or property of things or persons
An integrated circuit that contains the entire central processing unit of a computer on a single chip.

Micropyle
Greek
mikros- small
-pyle gate
Small opening at one end of an embryo sac.

Microscope
Greek
mikros- small
-skopein to view, examine
An optical instrument that uses a lens or a combination of lenses to produce magnified images of small objects.

Microspheres
Greek
mikros- small
-sphaera ball
Structures composed only of protein that have many properties of a cell.

Microtubules
Greek/Latin
mikros- small
-tubus- pipe
-ule little, small

Small hollow cylinders about 25 nm in diameter and 0.2–25 m in length.

Microvilli

Latin/Greek
mikros- small
-villus shaggy hair
Tiny hairlike folds in the plasma membrane that extend from the surface of many absorptive or secretory cells.

Microvolt

Greek
mikros- small
-volt electric potential
Small electric potential (one millionth of a volt).

Microwave

Greek/English
mikros- small
-waven undulating, wavy
Electromagnetic radiation of frequency 10^{10}–10^{12} Hz.

Micturation

Latin
mictum- to make water
-ion state, process, or quality of
The act or process of urinating.

Migration

Latin
migrans- to roam, wander, change places
-ion state, process, or quality of
The process of moving from one place to another.

Mimicry

Greek
mimikos- imitator or mimic
-y place for an activity; condition, state
A method of camouflage used in nature by an organism that involves the blending and concealment of one's identity by the effective use of color or shading.

Mineral

French
miniere- mine
-al of the kind of, pertaining to, having the form or character of
A naturally occurring, homogeneous inorganic solid substance having a definite chemical composition and characteristic crystalline structure, color, and hardness.

Mimicry in Nature

The process of natural selection has created some incredible relationships in nature. Members of all species seek the survival of their kind. Both prey and predator are subjected to environmental stresses on their numbers that can limit their growth and ultimately threaten their survival. This is a constant. Their abilities to adapt to changes, to modify their behaviors, and to compete with others for common resources such as food and water are continuously challenged in nature. But the amazing story is the process and randomness of natural selection. This selective process is not a willful or predetermined direction of genetic change, but rather the result of chance mutations over extended periods of time. It is the forces of nature that choose certain sets of phenotypes and eliminate others.

Consider the use of mimicry as a selective process. There are several varieties of mimicry, and all of them capitalize on characteristics that have sustained a population's growth in a given area. Batesian mimicry is the best known. This strategy is defined by a model species that possesses some sort of protective feature, such as a stinger, spines, or a toxin, and a species mimicking the model that does not. Batesian mimicry is exemplified by the American coral snake and the common milk or king snake. The coral snake is a venomous species with a very powerful poison, whereas the milk snake or king snake is not at all venomous. Yet the physical resemblance—the phenotype—is so striking that predators, including most humans, avoid the harmless snake. These snakes are marked with alternating yellow, red, and black bands. It is the arrangement of the bands that is the giveaway. The saying "Red against yellow: kill a fellow. Red against black: friend to Jack" is well known among Boy Scouts and outdoorsmen. There is little doubt that Batesian mimicry has allowed king snakes to flourish in the United States.

Miocene
Greek
meion- less
-kainos recent
An epoch of the Upper Tertiary period, spanning the time between 23.8 and 5.3 million years ago.

Miscible
Latin
miscere- to mix
-ible capable
Capable of undergoing mixing or blending.

Miticide
Latin
miti- mite
-cide (caedere) to cut, kill, hack at, or strike
A type of pesticide that kills mites that live on plants, livestock, and people.

Mitochondrion
Greek
mitos- warp thread
-khondro- granule, cartilage
-ion state, process, or quality of
Membranous organelle in which aerobic respiration continues and produces ATP molecules.

Mitogen
Greek
mit(os)- a thread
-gen- to give birth, kind, produce
Any substance or agent that stimulates mitotic cell division.

Mitosis
Greek/Latin
mitos- warp thread
-osis action, process, state, condition
The process in cell division by which the nucleus divides.

Mixture
Latin
miscere- to mix
-ure act, process, condition
The act of combining; any combination of materials that can be separated by ordinary physical means.

Mode
Latin
modus manner
In statistics, the average representing the sample value that occurs the most times; that which occurs most frequently in a series of observations.

Model
Latin
modulus small measure
A simplified version of a physical system that would be too complicated to analyze in full detail.

Molarity
German
mole- the amount of a substance containing Avogadro's number of units
-ar- relating to or resembling
-ity state of, quality of
The molar concentration of a solution.

Mole
German
molekulargewient molecular weight
Quantity of a substance that has a mass in grams numerically equal to its formula mass.

Molecule
Latin
moles- mass
-ule little, small
The smallest particle of a substance that retains all the properties of the substance and is composed of one or more atoms.

Molluscicide
Latin
mollusca- soft-bodied and prominent shell
-cide (caedere) to cut, kill, hack at, or strike
A type of pesticide that kills snails and slugs.

Mollusk
Latin
mollis- soft
molluscus thin-shelled
Phylum of animals having a soft, unsegmented body.

Moment
Latin
movere to move
The product of a quantity and its perpendicular distance from a reference point.

Momentum
Latin
movimentum to move
A measure of the motion of a body equal to the product of its mass and velocity.

Monoacid
Latin
mono- one, single, alone
-acere to be sour
An acid having one replaceable hydrogen atom.

Monoamine
Middle English
mono- one, single, alone
-amine any of a group of organic compounds

derived from ammonia by the replacement of one or more hydrogen atoms by a hydrocarbon radical An amine compound containing one amino group.

Monobasic
Latin
mono- one, single, alone
-base- basis
-ic (ikos) relating to or having some characteristic of
Having only one hydrogen ion to donate to a base in an acid-base reaction.

Monocotyledon
Greek
mono- one, alone, single
-kotuledon a kind of plant, a seed leaf, a hollow or cup-shaped object
Any of a class or subclass (Liliopsida or Monocotyledoneae) of chiefly herbaceous seed plants having an embryo with a single cotyledon, usually parallel-veined leaves, and floral organs arranged in cycles of three.

Monocular
Greek/Latin
mono- one, single, alone
-oculus eye
Of or pertaining to a single eye.

Monoecious
Greek
mono- one, single, alone
-oikos house
Having male and female sex organs on the same organism.

Monogamy
Greek
mono- one, single, alone
-gamos marriage
The condition of having a single mate at any one time.

Monohybrid
Greek
mono- one, single, alone
-hybrida mixed offspring
Pertaining to or describing an individual, organism, or strain that is heterozygous for the single trait or gene locus under consideration.

Monohydrate
Middle English
mono- one, single, alone
-hydr- water
-ate characterized by having
A crystalline compound that contains one molecule of water.

Monolayer
Middle English
mono- one, single, alone
-lay- to place in or bring to a particular state or position
-er one that performs an action
A film or layer of a compound one molecule thick.

Monomer
Greek
mono- one, single, alone
-meros a part, division
Small, individual molecule that forms a polymer.

Mononucleosis
Latin
mono- one, single, alone
-nucula- little nut, nucleus
-osis abnormal condition
A disease marked by extreme fatigue, high fever, and swollen lymph nodes, caused by an abnormally large number of white blood cells with single nuclei in the bloodstream.

Monothermia
Greek
mono- one, single, alone
-thermos- combining form of "hot" (heat)
-ia names of diseases, place names, or Latinizing plurals
A condition in which the temperature of the body remains the same throughout the day.

Monothetic
Greek
mono- one, single, alone
-thetikos- fit for placing
-ic (ikos) relating to or having some characteristic of
Denotes a taxonomic group classified on the basis of a single character, as opposed to polythetic.

Monotocous
Greek
mono- one, single, alone
-toco- childbirth, delivery, labor
-ous full of, having the quality of, relating to
Giving birth to but one offspring at a time.

Monotreme
Greek
mono- one, single, alone
-trema hole, perforation
The order of egg-laying (oviparous) mammals, including the duck-billed platypus and spiny anteater.

The Great Library of Alexandria

It can be said the Great Library of Alexandria (Egypt) was the best-known and one of the foremost libraries of the ancient world. Build by King Ptolemy II (309–246 BC) near where the temple of Muses (i.e., museum, from the word *musaeum*) once stood, this structure is now little more than a ruinous sublevel. But imagine an edifice so large that it contained an ornate main hall and ten great halls, each with armaria (i.e., wooden chests) containing thousands of handwritten papyrus scrolls from all points of the known world. Every one of the great halls was dedicated to a specific academic discipline. Scholars met, taught, and studied in an enlightened environment where knowledge and learning flourished.

Following the conquest of Egypt by Alexander the Great, the Greeks along with the Egyptians built this library as a seat where quite possibly all knowledge from the beginning of the world to the current time was archived and used by many of the most influential scientists, mathematicians, philosophers and artists. This massive repository housed

the compositions of philosophers Aristotle and Plato; the ancient Greek playwrights Sophocles and Euripides; the father of medicine, Hippocrates; the father of geometry, Euclid; and many other brilliant men, such as the legendary astronomer Aristarchus of Samos, who, in a missing manuscript, hypothesized a heliocentric solar system—that is, with the sun at the center and the planets, including earth, revolving around it. The manuscripts of one of the greatest mathematicians in history, Archimedes—"On the Equilibrium of Planes," explaining the laws of levers, and "On Floating Bodies," explaining the law of equilibrium of fluids—were also stored in the great library.

Men were sent to distant shores to copy manuscripts for the library. Ships were stopped at the port of Alexandria and searched for written works that could be borrowed and copied. The originals were kept in the library and copies were returned to the owners. We can only guess at how much scientific and mathematical knowledge had to be rediscovered because of the destruction of the library.

Historians dispute the who and when of the destruction of the Library of Alexandria. Julius Caesar had the port of Alexandria burned ca. 48 BC when he occupied the city. Scholars contend that that was a significant, but not a fatal, blow to the library. It is estimated that over 70,000 scrolls were destroyed by Caesar that day. However, many thousands of scrolls had been moved in anticipation of Caesar's conquest.

Some argue that Christian zealots in the fourth century destroyed the manuscripts, but not the library, because of the pagan teaching and learning that took place within its walls. Others say that the complete destruction of the library occurred at the hands of Muslims under the command of the Caliph Omar ca. AD 683, but this theory is discounted by most.

An inscription dedicated to Tiberius Claudius Babillus of Rome (d. AD 56) found at the Library of Alexandria supports the existence of the library after the time of Julius Caesar.

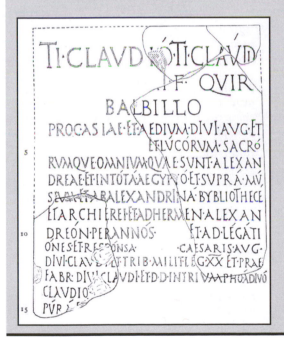

Monotrichous

Greek

mono- one, single, alone

-trich- hair

-ous full of, having the quality of, relating to

Having a single polar flagellum; said of a bacterial cell.

Monotropic

Greek

mono- one, single, alone

-trope- bend, curve, turn, a turning; response to stimulus

-ic (ikos) relating to or having some characteristic of

Affecting only one particular kind of bacterium, virus, or tissue; a narrowing of attention where an individual focuses on one entity.

Monsoon
Dutch (from Portugese)/Arabic
mawsim season
A wind system that influences large climatic regions and reverses direction seasonally.

Morainic
French
morena- mound of earth
-ic (ikos) relating to or having some characteristic of
Of or relating to an accumulation of boulders, stones, or other debris carried and deposited by a glacier.

Morphine
Latin
morph- shape, form, figure, or appearance
-ine a chemical substance
An opiate extract used in medicine to alleviate severe pain.

Morphogen
Greek
morph- shape, form, figure, or appearance
-gen to give birth, kind, produce
A class of substances that is said to be present in the embryo and that controls growth patterns.

Morphogenesis
Greek
morph- shape, form, figure, or appearance
-gen- to give birth, kind, produce
-sis action, process, state, condition
Formation of the structure of an organism or part; differentiation and growth of tissues and organs during development.

Morphology
Greek
morph- shape, form, figure, or appearance
-logy (logos) used in the names of sciences or bodies of knowledge
The study of the physical structures of organisms, in particular the soft tissues.

Mosaic
Greek
mouseion- shrine of the muses
An organism or part that is composed of two or more genetically distinct tissues, owing to experimental manipulation or to a faulty distribution of genetic material during mitosis.

Motion
Latin
movere- to move
-ion state, process, or quality of

An act, process, or instance of changing position.

Mucus
Latin
mucus mucus
A protective lubricant consisting of mucin, water, salts, and cells. This viscous fluid is secreted to protect cells, membranes, and various internal linings.

Multicellular
Latin
multus- much, many
-cella- chamber
-ar relating to or resembling
Consisting of many cells.

Muscle
Latin
mus mouse
Contractile tissue used to propel, move, and protect the body.

Museum
Greek
mouseion shrine of muses
An edifice or institution where cultural, scientific, historical, and contemporary artifacts, documents, and exhibits are retained for study and enjoyment.

Mutation
Latin
mut- change, changeable
-ion state, process, or quality of
A relatively permanent change in hereditary material, involving either a physical change in chromosome relations or a biochemical change in the codons that make up genes.

Mutualism
Latin
mutuus- borrowed or exchanged
-ism state or condition, quality
Association between organisms of two different species in which each member benefits.

Myalgia
Greek
myo- muscle
-algia pain, sense of pain; painful; hurting
Muscle pain.

Mycelium
Latin/Greek
myco- fungus
-helos- wart, nail, stud, corn
-ium quality or relationship
A mass of interwoven filamentous "threads" that make up the vegetative part of a fungus.

Mycology
Greek
myco- fungus
-logy (logos) used in the names of sciences or bodies of knowledge
The branch of botany that deals with fungi.

Mycorrhiza
Greek
myco- (mukes) fungi
-rhiza root
Mutualistic relationship between fungi and plants.

Myelin
Greek
myel- (muelos) bone marrow
-in protein or derived from a protein
A white fatty (lipid and lipoprotein) substance that is found in the medulla of long bones and also forms the insular layer of axons.

Myelodysplasia
Greek
myel- (muelos) bone marrow
-dys- painful, difficult, disordered, impaired, defective, ill
-plasia (plassein) something molded; to mold
Abnormal or defective (poor or bad) formation of the spinal cord.

Myocardium
Greek
myo- muscle
-kard- heart, pertaining to the heart
-ium quality or relationship
Specialized muscular tissue of the heart.

Myocyte
Greek
myo- muscle
-cyte (kutos) sac or bladder that contains fluid
Contractile cell (pinacocyte) in sponges.

Myofibril
Greek
myo- muscle
-fibrilla small fiber
Small part of a muscle fiber.

Myoglobin
Greek
myo- muscle
-globus- globular mass
-in protein or derived from a protein
Globular protein closely related to hemoglobin and located in the vertebrate muscle.

Myomere
Greek
myo- muscle
-meros part
A muscle segment of successive segmental trunk musculature.

Myometrium
Greek
myo- muscle
-metra- uterus
-ium quality or relationship
The smooth muscular layer lining the female uterus.

Myonecrosis
Greek
myo- muscle
-necro- death
-sis action, process, state, condition
Death of muscle tissue.

Myopia
Greek
muein- close to the eyes
-ops eye, optic
The condition of nearsightedness, where distant objects appear blurred.

Myosin
Greek
myo- muscle
-in protein or derived from a protein
Protein made up of a chain of polypeptides that forms filaments in smooth muscle fibrils.

Myotome
Greek
myo- muscle
-tomos (temnein) to cut, incise, section
A voluntary muscle segment in cephalochordates and vertebrates; that part of a somite destined to form muscles; the muscle group innervated by a single spinal nerve.

N

Nadir
Arabic
nazara to watch or see
The point of the celestial sphere directly under the observer; the opposite of zenith.

Naphtha
Greek
naphtha a flammable liquid issuing from the earth
A class of several volatile and flammable liquid mixtures of hydrocarbons that are distilled from petroleum, coal tar, and/or natural gases.

Nasal
Latin
nas- nose
-al of the kind of, pertaining to, having the form or character of
Of, in, or relating to the nose.

Nascent
Latin
nasc- born
-escent becoming
In the act of being formed, coming into existence, forming.

Nasopharynx
Latin
nasus- nose
-pharunx throat
The part of the pharynx above the soft palate that is continuous with the nasal passages.

Natural
Latin
natura- nature
-al of the kind of, pertaining to, having the form or character of
Of or pertaining to nature; that which occurs by chance or within the framework of natural design.

Nausea
Greek
nausie seasickness
A feeling of sickness in the stomach characterized by an urge to vomit.

Navel
Old English
nafela central point
The notch on the surface of the abdomen where the umbilical cord is attached during gestation.

Nebula
Latin
nebula cloud or mist
A diffuse mass of interstellar dust or gas or both, visible as luminous patches or areas of darkness depending on the way the mass absorbs or reflects incident radiation.

Necrobiosis
Greek
necro- death
-bios- life, living organisms, or tissue
-sis action, process, state, condition
The degeneration and death of the body's cells from natural processes.

Necrocoenosis
Greek
necro- death
-koinos- shared
-sis action, process, state, condition
An assemblage of dead organisms

Necrophagia
Greek
necro- death
-phagos (phagein) to eat, eating
Feeding on the flesh of dead animals.

Nectobenthos
Greek
necto- swim
-benthos deep; the fauna and flora of the bottom of the sea
Swimming off the seabed.

Nektonic
Greek
nekto- swimming
-ic (ikos) relating to or having some characteristic of
Describes numerous groups of marine and freshwater organisms capable of swimming against strong currents; these groups range from plankton to whales.

Nematic
Greek
nemat- thread, that which is spun
-ic (ikos) relating to or having some characteristic of
Refers to liquid crystals that have molecules arranged in loosely parallel lines.

Nematicide
Greek
nemat- thread, that which is spun
-cide (caedere) to cut, kill, hack at, or strike
A type of pesticide that kills nematodes (microscopic wormlike organisms that live in soil and cause damage to food crops).

Nematocyst
Greek
nemat- thread, that which is spun
-cyst (kustis) sac or bladder that contains fluid
Barbed harpoon within a cnidocyte of a cnidarian that is used to spear prey.

Nematoda
Greek
nemat- thread, that which is spun
-oeid shape, form, resembling
An order of worms having long, round, and generally smooth bodies.

Neon
Greek
neon new

A rare element that is a colorless, odorless, inert gas and that forms a very small part of the air.

Neoplasia
Greek
neos- new, recent
-plas- something made, molded, or formed
-ia names of diseases, place names, or Latinizing plurals
The transformation of a cell into a cancer cell.

Neoplasm
Greek
neos- new, recent
-plastos (plassein) something molded; to mold
An abnormal growth of new tissue in plants or animals; a tumor.

Neopterygian
Greek
neos- new, recent
-pteryx- fin
-ia names of diseases, place names, or Latinizing plurals
Any of a large group of bony fishes that includes most modern species.

Neoteny
Greek
neos- new, recent
-teinein to extend
An evolutionary process by which an organism produces a descendant that reaches sexual maturity while retaining a morphology characteristic of the pre-adult or larval stage of an ancestor.

Neotropical
Greek
neos- new, recent
-tropikos the tropics
Of, pertaining to, or designating a zoogeographical realm that includes Central and South America and the adjacent islands.

Nephelometer
Greek
nephele- cloud
-meter (metron) instrument or means of measuring; to measure
An instrument that determines the concentration of suspended matter in a liquid dispersion by measuring the amount of light that is scattered by the dispersion.

Nephric
Greek
nephros- kidney
-ic (ikos) relating to or having some characteristic of
Relating to or connected with a kidney.

Nephridium
Greek
nephros- kidney
-id state, condition; having, being, pertaining to
-ium quality or relationship
A tubular, glandular excretory organ characteristic of various coelomate invertebrates.

Nephritis
Greek
nephros- kidney
-itis inflammation, burning sensation
A variety of diseases causing chronic or acute inflammation of the kidneys.

Nephrolithotomy
Greek
nephros- kidney
-lithso- stone, rock
-tomos (temnein) to cut, incise, section
Incision made into the kidney for removal of stones.

Nephrology
Greek
nephros- kidney
-logy (logos) used in the names of sciences or bodies of knowledge
The science that deals with the kidneys, especially their functions or diseases.

Nephropexy
Greek
nephros- kidney
-pexy fixing of a specified part; attaching to, a fastening
Surgical fixation of a floating or mobile kidney.

Nephrosis
Greek
nephros- kidney
-sis action, process, state, condition
A noninflammatory disease of the kidneys that chiefly affects the function of the nephrons.

Nephrostome
Greek
nephros- kidney
-stoma mouth
Ciliated, funnel-shaped opening of a nephridium.

Neuralgia
Greek
neur- nerve, cord
nervus- sinew, tendon
-algia pain, sense of pain; painful, hurting
Acute pain radiating along the course of one or more nerves.

Neurilemma
Greek
neur- nerve, cord
nervus- sinew, tendon
-eilema veil, sheath
A very delicate sheathlike covering of a nerve fiber.

Neurilemmitis
Greek
neur- nerve, cord
nervus- sinew, tendon
-eilema- veil, sheath
-itis inflammation, burning sensation
Inflammation of the neurilemma.

Neurilemmoma
Greek
neur- nerve, cord
nervus- sinew, tendon
-eilema- veil, sheath
-oma tumor
Tumor of the peripheral nerve.

Neurilemmosarcoma
Greek
neur- nerve, cord
nervus- sinew, tendon
-eilema- veil, sheath
-sarko- flesh, meat
-oma tumor
A malignant neurilemma.

Neuroglia
Greek
neur- nerve, cord
nervus- sinew, tendon
-glia glue
Tissue supporting and filling the spaces between the nerve cells of the central nervous system.

Neurology
Greek
neur- nerve, cord
nervus- sinew, tendon
-logy (logos) used in the names of sciences or bodies of knowledge
Branch of science that deals with the study of the nervous system.

Neuromast
Greek
neur- nerve, cord
nervus- sinew, tendon
-mastos knoll, breast
Cluster of sense cells on or near the surface of a fish or amphibian that is sensitive to vibratory stimuli and to water current.

Neuron
Greek/Latin
neur- nerve, cord
nervus- sinew, tendon
-on a particle
A cell in the nervous system that is specialized to conduct nerve impulses, allowing different parts of the body to communicate.

Neuropeptide
Greek
neur- nerve, cord
nervus- sinew, tendon
-peptos- digestion, able to digest
-ide group of related chemical compounds
Any of various short-chain peptides found in brain tissue, such as endorphins.

Neuropodium
Greek
neur- nerve, cord
nervus- sinew, tendon
-podos foot
Lobe of the parapodium nearer the ventral side in polychaete annelids.

Neuroptera
Greek
neur- nerve, cord
nervus- sinew, tendon
-ptera feather, wing
Insect order for dobsonflies, ant lions, and lacewings, having four net-veined wings.

Neurotoxin
Greek
neur- nerve, cord
nervus- sinew, tendon
-tox- poison
-in protein or derived from a protein
A toxin that can damage nerve tissue.

Neurotransmitter
Greek/Latin
neur- nerve, cord
nervus- sinew, tendon
-trans- across
-mittere to send
Chemical substance released from the end of a neuron during the propagation of a nerve impulse, in order to transmit or pass a signal to another nerve cell.

Neurotrophic
Greek
neur- nerve, cord
nervus- sinew, tendon
-trophos- (trophein) to nourish, food, nutrition; development
-ic (ikos) relating to or having some characteristic of

Relating to the nutrition and metabolism of tissues under the influence of nerves.

Neutral
Greek
neutr- neither one nor the other
-al of the kind of, pertaining to, having the form or character of
In chemistry, a solution that is neither acidic nor basic, having a pH of 7.0.

Neutralization
Greek
neutr- neither one nor the other
-ation state, process, or quality of
In chemistry, the process of combining an acid and a base, thus canceling the properties of both and producing a salt and water.

Neutron
Greek
neutr- neither one nor the other
-on a particle
An uncharged elementary particle that has a mass nearly equal to that of the proton and is present in all known atomic nuclei except for the hydrogen nucleus.

Neutrophil
Greek
neutr- neither one nor the other
-phile one who loves or has a strong affinity or preference for
An abundant type of granular white blood cell that is highly destructive of microorganisms; it can be stained readily by neutral dyes.

Niche
Middle French
nicher to nest
The ecological role of an organism in a community, especially in regard to food consumption.

Nimbus
Latin
nimbus cloud
 Low, gray rain clouds.

Nocturnal
Latin
nocturnes- night
-al of the kind of, pertaining to, having the form or character of
Relating to, pertaining to, or occurring at night.

Nodule
Latin
nodus knot
-ulus small one
A small, knoblike outgrowth, such as those found on the roots of many leguminous plants.

Nomenclature

Latin

nom- (nemein) to dictate the laws of; knowledge; usage; order

-calator servant, crier

A system of names used in an art or science; the procedure of assigning names to kinds and groups of organisms in a taxonomic classification.

Nondisjunction

Latin

non- not, lack of

-jungere to join

The failure of paired chromosomes to separate during cell mitosis.

Nonideal

Greek

non- not, lack of

idea- a plan, scheme, notion, or method

-al of the kind of, pertaining to, having the form or character of

Pertains to a gas described by an equation of state of the form $pV = znRT$, where z is the gas deviation factor, which depends on pressure, temperature, and gas composition.

Nonpolar

Greek

non- not, lack of

-polos- either of two oppositely charged terminals, axis, sky

-ar relating to or resembling

Refers to a substance that does not ionize when combined with water.

Nonvascular

Latin

non- not, lack of

-vasculum- vessel

-ar relating to or resembling

Lacking a vascular system for the transport of nutrients throughout a plant.

Nonvolatile

Latin

non- not, lack of

-volare- to fly

-ile changing, ability, suitable, tending to

Pertains to that which does not readily evaporate at room temperature and pressure.

Noradrenaline

Latin

nor- anti or not

ad- to, a direction toward, addition to, near

-ren- the kidneys

-al of the kind of, pertaining to, having the form or character of

-ine a chemical substance

A hormone that acts directly on specific receptors to stimulate the sympathetic nervous system.

Norepinephrine

Greek

nor- anti or not

epi- above, over, on, upon

-nephros- kidneys

-ine a chemical substance

An endogenous adrenal hormone and synthetic adrenergic vasoconstrictor; this hormone constricts blood vessels and raises blood pressure.

Normal

Latin

norma- carpenter's square

-al of the kind of, pertaining to, having the form or character of

A perpendicular, especially a perpendicular to a line tangent, to a plane curve, to a plane tangent, or to a space curve.

Notochord

Greek

noton- back

-khorde gut, string of a musical instrument

A flexible rodlike structure that forms the main support of the body in the lowest chordates, such as the lancelet; a primitive backbone.

Notopodium

Greek

noton- back

-podos- foot

-ium quality or relationship

Lobe of a parapodium nearest the dorsal side in polychaete annelids.

Nucleic (acids)

Latin

nucula- kernel, little nut

-ic (ikos) relating to or having some characteristic of

A group of very large organic compounds important to the synthesis of protein molecules within cells. DNA and RNA are the two most widely known nucleic acids.

Nucleolus

Latin

nucula- kernel, little nut

-lus thing

A small, typically round granular body composed of protein and RNA, and found in the nucleus of a cell. It is usually associated with a specific chromosomal site and involved in ribosomal RNA synthesis and in the formation of ribosomes.

The Einstein-Szilard Letter

Months after the discovery of uranium fission in 1939, a Hungarian-born Jewish American physicist named Leo Szilard grew very concerned about the skepticism of American scientists that atomic energy from fission could be used for much of anything, let alone an atomic bomb. His fear was compounded by the fact that he and others believed Nazi Germany was working on a program to develop atomic weaponry. His suspicions were aroused by the discontinuation of uranium ore sales from Nazi-occupied Czechoslovakia.

If he was to persuade the Americans to begin a program of their own before it was too late, he had to convince President Roosevelt himself. Szilard sought the help of perhaps the best-known scientist in the world, Albert Einstein. Szilard, like Einstein, had fled Nazi Germany and come to America.

Szilard drafted a letter and took it to Einstein, who signed it and agreed to have it delivered to the president. Einstein was a pacifist, but he knew that if the Nazis had sole possession of such a weapon, it would mean defeat for the Allies in the coming war.

In the Einstein-Szilard letter, the scientists contended

This new phenomenon would also lead to the construction of bombs, and it is conceivable—though much less certain—that extremely powerful bombs of a new type may thus be constructed. A single bomb of this type, carried by boat and exploded in a port, might very well destroy the whole port together with some of the surrounding territory. However, such bombs might very well prove to be too heavy for transportation by air.

In the letter reprinted below, President Roosevelt gives his response.

This newly appointed "Uranium Board" had a limited scope of action and an extremely limited budget. Little to no action was taken toward the development of the atomic bomb until December 6, 1941, the day before the attack on Pearl Harbor by the Japanese. It was then that a large-scale research effort called the Manhattan Project began the process ultimately leading to the development of the atomic bomb dropped on Hiroshima, Japan, in August 1945.

Nucleonics
Latin
nucula- kernel, little nut
-ic (ikos) relating to or having some characteristic of
The science that deals with the study of the nucleus of atoms.

Nucleophile
Latin
nucula- kernel, little nut
-phile one who loves or has a strong affinity or preference for
A chemical compound or group that tends to donate or share electrons.

Nucleoplasm
Latin/Greek
nucula- kernel, little nut
-plasm (plassein) to mold or form cells or tissues

Protoplasm of a nucleus, as distinguished from cytoplasm.

Nucleosome
Latin/Greek
nucula- kernel, little nut
-soma (somatiko) body
Any one of the repeating nucleoprotein units consisting of histones forming a complex with DNA.

Nucleotide
Latin
nucula- kernel, little nut
-ide nonmetal radical
Chemical compounds consisting of a heterocyclic base combined with a sugar and one or more phosphate groups to form the basic structural units of DNA and RNA.

Nucleus
Latin
nucula- kernel, little nut
-us thing
In biology, a large, membrane-bound structure within a living cell, containing the cell's hereditary material and controlling its metabolism, growth, and reproduction. In chemistry, the positively charged central portion of an atom that comprises nearly all of the atomic mass and that consists of protons and neutrons—except in hydrogen, which consists of one proton only. In astronomy, the compact central core of a galaxy, often containing powerful radio, x-ray, and infrared sources.

Nutrient
Latin
nutrire- to suckle, nourish
-ent causing an action, being in a specific state, within
A source of nourishment or food.

Nyctalopia
Greek
nukt- night
-alaos- blind
-opia sight, eye
Night blindness.

Nyctanthous
Greek
nukt- night
-anthous flower
Describes plants that bloom or flower in the evening, such as jasmine.

O

Observation
Latin
ob- toward, against, before
-serv- to serve
-ation action, process, state, or condition
Any use of the senses to gather information.

Obstetrics
Latin/Greek
ob- toward, against, before
-statos- standing, stay; make firm, fixed, balanced
-ic (ikos) relating to or having some characteristic of
The branch of medicine that deals with the care of women during pregnancy, childbirth, and the recuperative period following delivery.

Occipital
Latin
ob- toward, against, before
-caput- head
-al of the kind of, pertaining to, having the form or character of
Of or pertaining to the back part of the skull; the occipital bone.

Occlude
Latin
occludere up close
To absorb and retain gases or other substances.

Occult
Latin
occulere to cover over
In medicine, a substance detectable only by microscopic examination.

Octahedron
Greek
octa- eight
-hedron face
A Platonic solid with eight faces.

Octet
Italian
oct- eight
-(du)et group
A set of eight valence electrons forming a stable configuration.

Octomerous
Greek
oct- eight
-meros part
Having eight parts; specifically, eightfold symmetry.

Oculomotor
Latin
oculus- eye, sight
-movere move
Moving or tending to move the eyeball.

Odometer
Greek
hodos- journey, way
-meter (metron) instrument or means of measuring; to measure
A mechanical or digital device used to record distance traveled.

Odonata
Greek
odontas toothed
An order of medium-to-large insects with elongated, slender abdomens; dragonflies and damselflies. Dragonflies hold wings horizontally when at rest, have thick bodies, and are active fliers. Damselflies hold wings vertically when at rest, have slender bodies, and are less agile in flight.

Odontoid
Greek
odontas- toothed
-oid (oeides) resembling, having the appearance of
Resembling a tooth; the odontoid process of the axis bone.

Oestrus
Greek
oistros having strong desire; anything that drives one mad; frenzy
The period during which the sexual desire and attractions of the female may be heightened, leading to copulation.

Olefin
French
oleum- oil
-fier form, cause to become
Any of a class of unsaturated open-chain hydrocarbons having the general formula C_nH_{2n}.

Olein
Latin
oleum- oil
-in natural chemical compound
An oily, yellow liquid occurring in animal and vegetable oil.

Olfaction
Latin
olfacere- smell
-ion state, process, or quality of
The process of smelling.

Oligocene
Greek
oligos- little, few
-kainos recent
An epoch of the Early Tertiary period, spanning the time between 33.7 and 23.8 million years ago.

Oligochaeta
Greek
oligos- little, few
-chaite long hair
Any of a class of hermaphrodite terrestrial or aquatic annelids (such as earthworms) that lack a specialized head.

Oligoclase
Greek
oligos- little, few
-klastos- break, break in pieces
-sis action, process, state, condition
Any of a class of common rocks forming series of triclinic feldspars.

Oligomer
Greek
oligos- little, few
-mer segment
A polymer that consists of two, three, or four monomers.

Oligosaccharide
Greek
oligos- little or few
-sakkhar- sugar
-ide nonmetal radical
A carbohydrate that consists of a relatively small number of monosaccharides.

Olivine
Latin (from Greek)
oliva- **(Latin)** color olive green
elaia- **(Greek)** olive green
-ine made of, resembling
A mineral silicate of iron and magnesium found in igneous and metamorphic rocks.

Ommatidium
Greek
omma- eye
-idium small
One of the optical units of the compound eye of arthropods and mollusks.

Omnivore
Latin
omnis- all
-vorare to devour
An organism that consumes a variety of plant and animal material.

Oncogene
Greek
onco- mass, bulk, swelling
-gen to give birth, kind, produce
A gene in which mutation induces neoplasia (cancer).

Oncosphere
Greek
onkinos- a hook
-sphaira ball
Rounded larva that is common to all cestodes and that bears hooks.

Ontogeny
Greek
onto- a being, individual; being, existence
-geny birth, descent, origin, creation, inception, beginning; race, sort, kind, class
The course of development of an individual organism. The history or science of the development of the individual being; embryology.

Oocyst
Greek
oion- egg
-cyst (kustis) sac or bladder that contains fluid
Cyst that forms around a zyogote of malaria and related organisms.

Oocyte
Greek
oion- egg
-cyte (kutos) sac or bladder that contains fluid
Stage in the formation of an ovum, just preceding the first meiotic division (primary oocyte) or just following the first meiotic division (secondary oocyte).

Oogenesis
Greek
oion- egg
-gen- to give birth, kind, produce
-sis action, process, state, condition
The formation, development, and maturation of an ovum.

Ookinete
Greek
oion- egg
-kinein to move
The motile zygote of malaria organisms.

Oolemma
Greek
oion- egg
-eilema veil, sheath
The plasma membrane of the oocyte.

Oology
Greek
oion- egg
-logy (logos) used in the names of sciences or bodies of knowledge
The branch of biology that deals with the study of eggs.

Oophoritis
Greek
oophor- ovary, egg
-itis inflammation, burning sensation
Inflammation of an ovary.

Ooze
Middle English

wose muddy ground
Soft mud or slime.

Opacity
Latin
opacus- shady
-ity state of, quality of
The quality or state of being opaque.

Opaque
Latin
opacus shady
Impenetrable by light; neither transparent or translucent.

Operator
Latin
operare- to work
-or a condition or property of things or persons
A genetic unit that regulates the transcription of structural genes in its operon.

Operculum
Latin
operire to cover
A lid or flap covering an aperture, such as the gill covers in some fish.

Operon
Latin
oper- operator
-on heredity unit
A unit of genetic material that functions in a coordinated manner by means of an operator, a promoter, and one or more structural genes that are transcribed together.

Ophthalmology
Greek
ophthalmos- eye; sight
-logy (logos) used in the names of sciences or bodies of knowledge
The branch of medicine that deals with the anatomy, functions, pathology, and treatment of the eye.

Ophthalmopathy
Greek
ophthalmos- eye; sight
-patheia disease; feeling, sensation, perception
The study of the diseases of the eye and associated tissue.

Opisthaptor
Greek
opistho- backward, behind, at the back, after, posterior
-haptein- to fasten
-or a condition or property of things or persons
The posterior attachment organ of a monogenetic trematode.

Opisthognathous
Greek
opistho- backward, behind, at the back, after, posterior
-gnathos jaw
With the head deflexed such that the mouthparts are directed posteriorly, as in the insect order Hemiptera.

Opsonin
Greek
opson- a relish
-in protein or derived from a protein
Type of antibody in blood serum that weakens bacteria and other foreign cells so that the phagocytes can destroy them more easily.

Optic
Greek
optikos- visable
-ic (ikos) relating to or having some characteristic of
Referring to vision or the science of optics or lenses.

Orbital
Latin
orbita- orbit
-al of the kind of, pertaining to, having the form or character of
Refers to the wave function of an electron in an atom or molecule.

Organ
Greek
organon- organized structure; pertaining to a particular body part with a specific function(s); tool, implement
The aggregation of various tissues into a specific structure designed to carry out some biological function within a multicellular organism.

Organelle
Greek/Latin
organon- organized structure; pertaining to a particular body part with a specific function(s); tool, implement
-elle diminutive
Specialized part of a cell; literally, a small organ that performs functions analogous to those of organs of multicellular animals.

Organic
Greek
organon- organized structure; pertaining to a particular body part with a specific function(s); tool, implement
-ic (ikos) relating to or having some characteristic of
Of or pertaining to compounds containing carbon.

Johannes Kepler

It had been well over 1500 years since the first and perhaps only major paradigm in science had swept the Western world. Now the paradigm was about to shift. A bold new group of thinkers had emerged in Europe to challenge the accepted theories and to lay the foundation for a more progressive approach to science (a newly coined word) and experimentation. The scientific revolution was about to begin.

Johannes Kepler, born in Germany on December 27, 1571, was one of the first to question contemporary thinking. He wrote, "Geometry existed before the Creation. It is co-eternal with the mind of God. . . . Geometry is God himself."

Even as a child, Kepler was gifted and outspoken. He studied religion, mathematics, and philosophy at a Protestant seminary school. In his relatively sequestered life, he pondered the relationship between God and the natural world. He looked for mathematical evidence of harmony between the eternal and the natural. One might even describe him as a patron of Pythagoras. For a time he believed in the Platonic solids as a framework for the orbits of the planets.

The number of known planets in Kepler's time was six. To Kepler, the nagging question was, why only six? Why not more? He struggled with the explanation of the distances between the planets according to Copernicus. He spent years trying to formulate a reasonable explanation of the data on planetary positions that he had obtained from Tycho Brahe. He wanted to develop an experimental approach to studying planetary design, but he needed baseline data. He brilliantly determined that by using the sun and the orbital period of Mars, he could produce data establishing that the orbital path of Mars was not circular. To Kepler, such disharmony was very unsettling, but he clearly demonstrated that the order and perfection of the heavens, as described by the Greeks, was more myth than fact.

Organism

Greek

organon- organized structure; pertaining to a particular body part with a specific function(s); tool, implement

-ism state or condition, quality

An individual living animal or plant able to carry on life functions through mutually dependent systems and organs.

Organogenesis

Greek/Latin

organon- organized structure; pertaining to a particular body part with a specific function(s); tool, implement

-gen- to give birth, kind, produce

-sis action, process, state, condition

The formation and development of the organs of living things.

Organosol

Greek

organon- organized structure; pertaining to a particular body part with a specific function(s); tool, implement

-ic (ikos) relating to or having some characteristic of

-ol chemical additive

A colloidal dispersion in which an organic dispersion medium is used.

Orientation

Latin

orient- to adjust

-ion state, process, or quality of

Change of position by organs, organelles, or organisms in response to external stimulus.

Orifice

Latin

or- mouth

-ficium a making, doing

An opening to a cavity or to a body; mouth.

Ornithodelphia

Greek

ornis- bird

-delphys- womb

-ia names of diseases, place names, or Latinizing plurals

Infraclass of monotreme mammals.

Ornithology

Greek

ornis- bird

-logy (logos) used in the names of sciences or bodies of knowledge

The branch of zoology dealing with the scientific study of birds and their structure, classification, habits, songs, and flight.

Orogeny

Greek/French

oros- mountain

-gen- to give birth, kind, produce

-y place for an activity; condition, state

The formation of mountains through plate tectonics.

Oropharynx

Greek

or- mouth

-pharynx cavity leading from the mouth and nasal passages to the larynx

The part of the pharynx that extends from the mouth to the larynx.

Orpiment

Latin

aurum- gold or yellow

-pigmentum pigment

A bright yellow mineral, arsenic trisulfide, that is used as a pigment.

Orthoclase

Greek

ortho- straight, true, correct, right

-klasis to break

A variety of feldspar, essentially potassium aluminum silicate, or $KAlSi_3O_8$, characterized by a monoclinic crystalline structure and found in igneous or granitic rock.

Orthogenesis

Greek

ortho- straight, true, correct, right

-gen- to give birth, kind, produce

-sis action, process, state, condition

The idea that the evolutionary path of a lineage can acquire a trend that carries it in a continuous direction; directional selection.

Orthopedics

Greek

ortho- straight, true, correct, right

-paideia- child rearing

-ic (ikos) relating to or having some characteristic of

The branch of medicine that deals with the prevention or correction of injuries or disorders of the skeletal system and associated muscles, joints, and ligaments.

Orthoptera

Greek

ortho- straight, true, correct, right

-ptera feather, wing

An order of mandibulate insects including grasshoppers, locusts, and cockroaches; insects with greatly enlarged hind legs with forewings modified into a tegmen.

Oscillate
Latin
os- mouth
-cillum to swing
To vary between alternate extremes, usually within a definable period of time.

Osculum
Latin
os- mouth
-culum diminutive, little
Excurrent opening in a sponge.

Osmiridium
English
osme- from the smell of osmium tetroxide
-irid- rainbow
-ium quality or relationship
A mineral that is a natural alloy of osmium and iridium, with small inclusions of platinum, rhodium, and other metals.

Osmium
Greek
osme- smell from the smell of osmium tetroxide
-ium quality or relationship
A hard metallic element found in small amounts in osmiridium and platinum ores.

Osmosis
Greek
osmos- thrust, push
-osis action, process, state, condition
Diffusion of fluid through a semipermeable membrane from a solution with a low solute concentration to a solution with a higher solute concentration, until there is an equal concentration of fluid on both sides of the membrane.

Osmotic
Greek
osmos- thrust, push
-ic (ikos) relating to or having some characteristic of
Relating to the diffusion of a fluid through a semipermeable member until there is equal concentration on both sides of the membrane.

Osmotroph
Greek
osmos- thrust, push
-trophos (trophein) to nourish, food, nutrition; development
A heterotrophic organism that absorbs dissolved nutrients.

Ossification
Latin
oss- bone
-ify- (ficus) make, or cause to become
-ion state, process, or quality of
The natural process of forming bone from soft tissue, including cartilage and membranous tissue.

Osteichthyes
Greek
osteon- bone
-ichthus fish
A class of fish having a skeleton composed of bone in addition to cartilage.

Osteoarthropathy
Greek
osteon- bone
-arthr- joint
-patheia disease, feeling, sensation, perception
A disorder affecting bones and joints.

Osteoblast
Greek
osteon- bone
-blastos bud, germ cell
Cells that help create bone by facilitating the deposit of minerals.

Osteoclast
Greek
osteon- bone
-klastos break, break in pieces
A large, multinucleate cell found in growing bone that reabsorbs bony tissue, as in the formation of canals and cavities.

Osteocyte
Greek
osteon- bone
-cyte (kutos) sac or bladder that contains fluid
A cell embedded in a bone.

Osteology
Greek
osteon- bone
-logy (logos) used in the names of sciences or bodies of knowledge
Part of anatomy dealing with the study of the structure, development, and function of bones.

Osteopathy
Greek
osteon- bone
-patheia disease, feeling, sensation, perception
Disease involving the bones.

Osteoporosis
Greek
osteon- bone
-poros- a passage
-sis action, process, state, condition
A disease in which the bones become porous.

Antoine Lavoisier

Antoine Lavoisier is considered by many to be the father of modern chemistry. That title, however, was not enough to save him from the guillotine in 1794. He was born in Paris, France, on August 26, 1743, to a family of wealth and privilege. Lavoisier never endeared himself to the public. He worked for a time as a tax collector in Paris. Clearly, he was in the wrong profession at the wrong time. Nothing he did scientifically could make up for the aristocratic persona Lavoisier projected in the earlier years of his life. Thus, when he made his final appeal to the judge in the French court, the judge's response was simply "the Revolution has no need of scientists." He was taken out and executed along with many others, including his father-in-law, who was executed right before him.

Antoine Lavoisier was a remarkable chemist. He was one of the first to quantify chemistry, that is, to assign numbers to chemicals and to chemical reactions. The law of conservation of matter was a direct result of Lavoisier's experiments. By carefully weighing both reactants and products, he demonstrated that the mass of the end products of a chemical reaction is equal to the mass of the reactants.

Prior to the work of Lavoisier, there had only been discussion of the possibility of the existence of compounds. By his clever quantification of chemical reactions, Lavoisier was able to prove that elements do, in fact, combine to form compounds. Lavoisier was the first to prove that water was a compound composed of the elements hydrogen and oxygen. He also demonstrated that the ratio of hydrogen to oxygen is 2 to 1. Lavoisier's *Elementary Treatise of Chemistry*, published in 1789, was considered by many to be the first chemistry textbook. It encapsulated in an integrated perspective a modern approach to chemistry and chemical analysis. In addition to creating a chemical nomenclature and discounting previously accepted chemical theories, such as the phlogiston theory of matter, he introduced in his writings a significant group of chemicals that could not be broken down further. Those chemicals are many of the elements we are familiar with today.

All this and more could not save him. His country was in turmoil, and the French Revolution turned even more violent in its latter stages. When Lavoisier was arrested and brought to court, no one stood in his defense. His peers and closest friends, who knew he was innocent of the serious charges brought against him, did nothing and said nothing. Everyone feared for their own lives. The terror that was the French Revolution struck such fear in the hearts of men that they allowed the innocent to go down with the guilty.

A very short year and a half later, the French government exonerated Lavoisier of all guilt. Too little and far too late.

Ostium

Latin
os- mouth
-ium quality or relationship
Name given to any small opening in an organism; mouthlike opening in organisms; one of the small porelike openings in sponges.

Otodynia

Greek
ot- ear; relationship to the ear
-dynia pain
Pain in the ear; earache.

Otolith

Greek
ot- ear; relationship to the ear
-lithos stone, rock
Calcerous concretions in the membranous labyrinth of the inner ear of lower vertebrates or in the auditory organ of certain vertebrates.

Outcrop

Old English
ut- away from the center or middle
-crop to appear on the surface
A portion of bedrock or other stratum protruding through the soil level.

Ovary

Latin
ovum- egg
-ary of, relating to, or connected with
The ovule-bearing lower part of a pistil that ripens into a fruit.

Ovicide

Latin
*ovum-*egg
-cide (caedere) to cut, kill, hack at, or strike
A type of pesticide that controls insect eggs through the application of low-sulfur petroleum oils to plants and animals.

Oviger

Greek
ovum- egg
-gerere to bear
Leg that carries eggs in pycnogonids.

Ovine

Latin
ov- sheep
-ine of or relating to
Refers to sheep.

Ovipositor

Latin
ovum- egg
-pos- to place
-or a condition or property of things or persons, person who does something
Organ of female insects through which eggs are laid.

Ovoviviparity

Latin
ovum- egg
-vivi- life, alive
-parity to bring forth, to bear, producing viable offspring, giving birth to
Retention of the developing fertilized egg within the mother; a form of viviparity in which there is no nutrition of hatched young.

Ovulation

Latin
ovum- egg
-ation action, process, state, or condition
The process of releasing the ovum from the ovary.

Ovule

Latin
ovum- egg
-ule little, small
A minute structure in seed plants that develops into a seed after fertilization.

Ovum

Latin
ov- egg
-um (**singular**) structure
-a (**plural**) structure
Plural *ova*; female gamete before fertilization.

Oxalate

French/Latin
oxal- a derivative of oxalic acid, found in plants
-ate meaning the salt or ester of the root acid C_2O_4, the ion of oxalic acid $Na_2C_2O_4$, salt of oxalic acid.

Oxidation

French
oxide- a binary compound of an element or a radical with oxygen
-ion state, process, or quality of
A reaction in which the atoms in an element lose electrons and the valence of the element is correspondingly increased (originally, this was considered to be the combination of a substance with oxygen).

Oxygen

Latin/Greek
oxus- acid, sharp
-gen to give birth, kind, produce
A nonmetallic element constituting 21% of the atmosphere by volume that occurs as a diatomic gas, O_2, and in many compounds such as water and iron ore.

P

Palate

Greek/Latin

pal- flat

-ate characterized by having

In mammals, the roof of the mouth. The bony front part is the hard palate, and the muscular rear part is the soft palate.

Paleoanthropology

Greek

palaois- ancient, old

-anthropo- human

-logy (logos) used in the names of sciences or bodies of knowledge

The study of fossils belonging to the genus *Homo* (e.g., *Homo erectus*).

Paleocene

Greek/Latin

palaois- ancient, old

-recens recent

The earliest epoch of the Tertiary period, spanning the time between 65 and 55.5 million years ago.

Paleontology

Greek

palaois- ancient, old

-ontos- having existed

-logy (logos) used in the names of sciences or bodies of knowledge

The study of the forms of life existing in prehistoric or geologic times, as represented by the fossils of plants, animals, and other organisms.

Paleozoic

Greek

palaois- ancient, old

-zoikos- of animals

-ic (ikos) relating to or having some characteristic of

The second oldest division of geologic time; an era of geologic time from the end of the Precambrian to the beginning of the Mesozoic.

Palpitate

Latin

palpare- to feel

-ate characterized by having

To beat rapidly, as the heart.

Pandemic

Greek

pan- all

-demos- the people

-ic (ikos) relating to or having some characteristic of

An epidemic over a large region.

Paracentesis

Greek

para- beyond

-cente- puncture

-sis action, process, state, condition

The process of aspirating a cavity.

Paradox

Greek

para- beyond

-doxa explanation

A seemingly contradictory statement that may nonetheless be true.

Paraffin
Latin
parum- little, not very
-affinis associated with
A member of the alkane series.

Parallax
Greek
para- beside; near; alongside
-allos other
The apparent change in the position of an object resulting from the change in the direction or position from which it is viewed.

Parallel
Greek
para- beside; near; alongside
-allos one another
Extending in the same direction; everywhere equidistant and not meeting.

Paralysis
Greek
para- beside; near; alongside
-luein- to release
-sis action, process, state, condition
The loss of either sensation or movement or both on a part of the body, usually as a result of injury.

Paramagnetic
Greek
para- beside; near; alongside
-magnes- stone from Magnesia (city in Asia Minor)
-ic (ikos) relating to or having some characteristic of
Relating to or being a substance in which an induced magnetic field is parallel and proportional to the magnetizing field, but is much weaker than in ferromagnetic materials.

Paramecium
Greek
para- beside; near; alongside
-mekos- length
-ium quality or relationship
Freshwater species of the genus *Paramecium* that is typically long and narrow, with an oral groove on the side.

Parasite
Greek
para- beside; near; alongside
-sitos- grain, food
-ite resident
An organism that grows, feeds, and is sheltered on or in a different organism while contributing nothing to the survival of its host.

Parasitism
Greek

para- beside; near; alongside
-sitos- grain, food
-ism state or condition, quality
The condition of an organism living in or on another organism at whose expense the parasite is maintained.

Parasitology
Greek
para- beside; near; alongside
-sitos- grain, food
-logy (logos) used in the names of sciences or bodies of knowledge
A branch of science that deals with parasites and parasitism.

Parathyroid
Greek
para- beside; near; alongside
-thureos- oblong shield; door
-oid (oeides) resembling, having the appearance of
Four small kidney-shaped glands located laterally and posteriorly to the thyroid glands in the neck; they secrete the parathyroid hormone.

Parenchyma
Greek
para- beside; near; alongside
-enchyma infusion
Least specialized of all plant cell or tissue types.

Parietal
Latin
pariet- wall
-al of the kind of, pertaining to, having the form or character of
In biology, refers to either the parietal bone of the skull or the forming of a wall of a body part or organ.

Parity
Latin
par- equal
-ity state of, quality of
An intrinsic symmetry property of subatomic particles that is characterized by the behavior of the wave function of such particles under reflection through the origin of spatial coordinates.

Parotid
Greek
par- by the side of, beside; associated, near
-id state, condition; having, being, pertaining to, tending to, inclined to
Pertaining to the salivary glands located on the side of the head near the ears.

Parotitis
Greek

par- by the side of, beside; associated, near
-itis inflammation
Inflammation of the parotid glands, as in mumps.

Parsec (Parallax- second)
Greek
para- beside; near; alongside
-allos- other
-sec (secundus) second
A distance at which an object will have a parallax of one second of arc; 3.258 light years or 1.918 Þ 10^{23} miles.

Parthenogenesis
Greek
parthenos- virgin
-gen- to give birth, kind, produce
-sis action, process, state, condition
A form of reproduction in which an unfertilized egg develops into a new individual, occurring commonly among insects and certain other arthropods.

Particle
Latin
particula part
Any of the basic units of matter and energy.

Pathogenic
Greek
pathos- suffering, disease
-gen- to give birth, kind, produce
-ic (ikos) relating to or having some characteristic of
Refers to an agent, typically a microbe that causes disease or suffering.

Pathology
Greek
pathos- suffering, disease
-logy (logos) used in the names of sciences or bodies of knowledge
The science of disease formation, processes, causes, and effects.

Pediatrics
Greek
paideia- child rearing
-iasthai to heal
The branch of medicine that deals with the care of infants and children and the treatment of their diseases.

Pedigree
French
ped- foot
-de grue of crane (resembling a crane's foot)
A diagram that traces a trait through several family generations.

Pedipalp
Latin
ped- foot
-palp, -palpi, -palpo to touch, stroke
One of the second pair of appendages near the mouth of a spider or other arachnid that are modified for various reproductive, predatory, or sensory functions.

Peduncle
Latin
ped- foot
-uncle little
A primary flower stalk, supporting either a cluster or a solitary flower.

Pelagic
Greek
pelagikos- (pelagos) sea
-ic (ikos) relating to or having some characteristic of
Of, relating to, or living in open oceans or seas rather than in waters adjacent to land or in inland waters.

Pellicle
Latin
pellicula husk
Thin, protective membrane in some protozoa.

Pelvis
Latin
pelvis basin
A basin-shaped cavity at the base of the axial skeleton formed by the fusion of six bones, the ileum, pubis, and the ischium.

Penetrometer
Latin
penetr- inner or inside
-meter (metron) instrument or means of measuring; to measure
An instrument designed to measure the density, compactness, and penetrability of a substance.

Penguin
Old Welsh
pen- white
-gwyn head
Any of various erect, short-legged, flightless aquatic birds (family Spheniscidae) of the Southern Hemisphere.

Penicillin
Latin
penicillus- brush
-in protein or derived from protein
Any of a group of broad-spectrum antibiotic drugs obtained from penicillium molds or produced synthetically; most active against gram-positive bacteria and used in the treatment of various infections and diseases.

Pentahedron
Greek
penta- five
-hedron face
A three-dimensional solid having five (plane) faces.

Pentamer
Greek
penta- five
-meros a part
A polymer consisting of five molecules.

Penumbra
Latin
paene- almost
-umbra shadow
The outer, almost darkened part of a shadow cast during an eclipse that lies between the completely darkened area and the fully lit area.

Peptide
English
pept(one)- digested
-ide group of related chemical compounds
Any of various natural compounds containing two or more amino acids linked by the carboxyl group of one amino acid and the amino group of another.

Peptize
Greek
pept(one)- digested
-ize to make, to treat, to do something with
To change a gel into a colloid solution form.

Percolate
Latin
per- through, across
-co- together, with
-late bear, carry
To cause a liquid to pass through spaces of a porous material.

Perennial
Latin
per- through, across
-annus- year
-al of the kind of, pertaining to, having the form or character of
Refers to that which lasts year after year; a perennial plant.

Pericardia
Greek
peri- around, about, enclosing
-kard- heart, pertaining to the heart
-ia names of diseases, place names, or Latinizing plurals

Thin, membranous, fluid-secreting sac in the area around the heart.

Pericarditis
Greek
peri- around, about, enclosing
-kard- heart, pertaining to the heart
-itis inflammation, burning sensation
Inflammation of the tissue surrounding the heart.

Pericycle
Greek
peri- around, about, enclosing
-kyklos circle, wheel, cycle
Thin tissue layer found in vascular plants; can produce lateral roots.

Peridotite
French
peridot- a yellowish green variety of olivine used as a gem
-ite minerals and fossils
Any of a group of igneous rocks composed mainly of olivine and various pyroxenes and having a granitelike texture.

Perigee
French (from Greek)
peri- around, about, enclosing
-ge earth, world
The point nearest the earth's center in the orbit of a moon or satellite.

Perihelion
Greek
peri- around, about, enclosing
-helios- sun
-ion state, process, or quality of
The point along an orbit of a planet at which the planet is closest to the sun.

Perimorph
Greek
peri- around, about, enclosing
-morph shape, form, figure, or appearance
A mineral that encloses a different mineral.

Perineum
Greek
peri- around, about, enclosing
-inan to excrete
In females, the area between the anus and the vagina.

Period
Greek
peri- around, about, enclosing
-hodos journey, way
The geological length of time.

Periodic
Greek
peri- around, about, enclosing
-hodos- journey, way
-ic (ikos) relating to or having some characteristic of
Having or marked by repeated cycles.

Perissodactyla
Greek
perissos- odd
-dactylos toe
Order of odd-toed mammals (horses, zebras).

Peristalsis
Greek
peri- around, about, enclosing
-stellein- to place
-sis action, process, state, condition
Muscular contractions of esophagus.

Peritoneum
Greek
peri- around, about, enclosing
-teinein to stretch
The membrane that lines the walls of the abdominal cavity.

Peritrichous
Greek
peri- around, about, enclosing
-tricho- made of hair
-ous full of, having the quality of, relating to
Pertains to having flagella all over a cell.

Permafrost
Latin/Middle English
permanere- to endure
-frost freeze; frozen
Permanently frozen subsoil continuous throughout the polar region.

Permeable
Latin
per- through
-meare- to glide
-able capable, be inclined to, tending to, given to
Capable of being penetrated by liquids or gases.

Peroxide
Latin
per- large or largest portion of an element
-oxy(s)- sharp, acid
-ide group of related chemical compounds
An oxide of an element or a radical that contains the greatest possible amount of oxygen, especially when there are oxygen atoms joined to each other.

Peroxisome
Latin/Greek
per- large or largest portion of an element

-oxy(s)- sharp, acid
-soma (somatiko) body
A cell organelle containing enzymes such as catalase and oxidase that catalyze the production and breakdown of hydrogen peroxide.

Pesticide
Latin
pesti- plague, contagion
-cide (caedere) to cut, kill, hack at, or strike
A chemical agent used to destroy pests.

Petal
Greek
petalon leaf
One of the often brightly colored parts of a flower immediately surrounding the reproductive organs.

Petrochemical
Greek
petros- a rock, fossil, or stone
-chemeia- alchemy
-al of the kind of, pertaining to, having the form or character of
A chemical derived from fossil fuels.

Petroleum
Latin
petros- a rock, fossil, or stone
-oleum oil
Oily, flammable liquid that occurs naturally in deposits, usually beneath the surface of the earth.

Petrology
Greek
petros- a rock, fossil, or stone
-logy (logos) used in the names of sciences or bodies of knowledge
Branch of geology that deals with the study of rocks, their mineral compositions, their textures, and their origins.

Phagocyte
Greek
phagos- (phagein) to eat, eating
-cyte (kutos) sac or bladder that contains fluid
White blood cells that destroy pathogens by surrounding and engulfing them.

Phagocytosis
Greek
phagos- (phagein) to eat, eating
-cyte- (kutos) sac or bladder that contains fluid
-sis action, process, state, condition
The process by which a cell absorbs or eats waste materials.

Phanerozoic

Greek

phainein- visible

-zoion living being

The most recent past geologic eon that includes the Cenozoic, Mesozoic, and Paleozoic eras.

Pharmacology

Greek

pharmac- drug, medicine, or poison

-logy (logos) used in the names of sciences or bodies of knowledge

The study of the properties of drugs and their effects on the body.

Pharyngotomy

Greek

pharyng- throat

-tomos (temnein) to cut, incise, section

An operation in which an incision is made into the pharynx to remove a tumor.

Pharynx

Greek

pharyng- throat

Passage between the esophagus and the cavities of the nose and mouth.

Phenocryst

Greek

phaino- showing, displaying

-krustallos ice, crystal, freeze, icelike

A conspicuous, usually large, crystal that is embedded in porphyritic igneous rock.

Phenol

Greek

phen- related to or derived from benzene

-ol chemical derivative

A caustic, poisonous, white crystalline compound derived from benzene and used in resins, plastics, and pharmaceuticals, as well as in dilute form as a disinfectant and antiseptic.

Phenology

Greek

phainein- to show, appear, display; making evident; literally, "to come"

-logy (logos) used in the names of sciences or bodies of knowledge

The seasonal life history of an insect population.

Phenomenon

Greek

phainomenon to appear

An observable event.

Phenotype

Greek

phainein- to show, appear, display; making evident; literally, "to come"

-typos mark

The complete observable characteristics of an organism or group including anatomic, physiologic, biochemical, and behavioral traits as determined by the interaction of genetic makeup and environmental factors.

Pheromone

Greek

pherein- to carry, bear, support; go

-(hor)mone to rouse, or set in motion

A chemical secreted by an animal, especially an insect, that influences the behavior or development of others of the same species and often functions as an attractant of the opposite sex.

Philodendrist

Greek

philos- love, fondness for, loving

-dendron- tree

-ist one who is engaged in

One who has a special fondness for trees.

Phlebitis

Greek

phleb- blood vessel, vein

-itis inflammation, burning sensation

The inflammation of a vein.

Phlebosclerosis

Greek

phleb- blood vessel, vein

-skleros- hard

-sis action, process, state, condition

Thickening or hardening of the walls of the veins.

Phloem

Greek

phloios bark

The food-conducting tissue of vascular plants.

Phosphorus

Greek

phos- light

-pherein to carry, bear, support; go

A highly reactive, poisonous, nonmetallic element found in safety matches and pyrotechnics.

Photochemical

Latin

photos- light, radiant energy

-alchymia- action of chemicals

-al of the kind of, pertaining to, having the form or character of

Refers to chemicals and other pollutants reacting in the presence of sunlight.

Photoelectric
Greek
photos- light, radiant energy
-elector- beaming sun
-ic relating to or having some characteristic of
Pertains to the ejection of an electron from a surface exposed to light.

Photometry
Greek
photos- light, radiant energy
-metria (metron) the process of measuring; to measure
The branch of science that deals with the measurement of light output.

Photon
Greek
photos- light, radiant energy
-on a particle
The smallest physical particle; it has no mass and no charge, and is electromagnetic energy.

Photopsin
Greek
photos- light, radiant energy
-opsis- sight, appearance
-in neutral chemical. protein derivative
The photoreceptor pigments found in the cone cells of the retina that are the basis of color vision.

Photoreceptor
Greek
photos- light, radiant energy
-recept- receiver
-or a condition or property of things or persons, person who does something
A group of nerve cells that are sensitive to light energy.

Photosensitive
Greek
photos- light, radiant energy
-sensus- senses
-ive performing an action
Refers to something that is easily irritated by light.

Photosphere
Greek
photos- light, radiant energy
-sphaira a globe shape, ball, sphere
The intensely bright gaseous outer layer of a star, especially of the sun.

Photosynthesis
Greek
photos- light, radiant energy
-synthe- formation by combination
-sis action, process, state, condition
The process by which carbon dioxide is converted into organic matter in the presence of the chlorophyll in plants and under the influence of light.

Phototropism
Greek
photos- light, radiant energy
-trope- bend, curve, turn, a turning; response to stimulus
-ism state or condition, quality
Adjustment in the direction and rate of plant growth in response to light.

Phycoerythrin
Greek
phukos- seaweed
-erythros red
A red phycobilin occurring especially in the cells of red algae.

Phyllotaxy
Greek
phullon- leaf
-taxi arrangement, order
The manner in which leaves are arranged with regard to the axis.

Phylogeny
Greek
phulon- race, class, tribe
-genes to give birth, kind, produce
Development and history of a species or higher taxonomic grouping of organisms.

Phylum
Greek
phulon- race, class, tribe
The chief category of taxonomic classifications, between kingdom and class, into which organisms of common descent that share a fundamental pattern of organization are grouped.

Physical
Greek
physica- physics
-al of the kind of, pertaining to, having the form or character of
In physics, a term used to refer to or identify material things. In biology, a term used to refer to or denote the body as opposed to the mind or spirit.

Ernest Rutherford

Ernest Rutherford is considered by many to be the father of nuclear physics. He was born Earnest Rutherford, the first Baron Rutherford of Nelson, in New Zealand on August 30, 1871. He died on August 19, 1937.

Rutherford became known for developing an experimental design demonstrating the scattering of nuclear (alpha) particles using gold foil. For a time, he studied at the University of Cambridge in England, where, during his investigations of wireless wave energy and radioactivity, he coined the terms *alpha*, *beta*, and *gamma rays*.

Rutherford moved to Canada and took a professorship in and chaired the Department of Physics at McGill University. There he developed an explanation for the constant rate of disintegration of radioactive atoms, ultimately leading to the term *half-life*. He went on to associate this process of atomic decay with a precise, clocklike action. By examining the half-life of radium and knowing that radium ultimately came from the degradation of uranium, Rutherford was able to speculate about the age of the earth. He placed the age at hundreds of millions of years—not exactly accurate or narrow in its scope, but it was a starting point that was picked up by scientists later on. For this work, he was awarded a Nobel Prize in Chemistry in 1908.

Rutherford began to feel left out of mainstream science at McGill, so he moved to Great Britain and was given the chair of the Department of Physics at the University of Manchester. Here he ultimately discovered the nature of the nuclei of atoms. He theorized about "neutrons" in the nuclei as being particles capable of countering the effects of positively charged protons and thus preventing the nucleus from breaking apart.

His pioneering work in nuclear physics was instrumental in the establishment of the Manhattan Project. During his work in nuclear science, Rutherford was quoted as saying, "The energy produced by breaking down the atom is a very poor kind of thing. Anyone who expects a source of power from the transformations of these atoms is talking moonshine."

He desperately wanted to avoid the development of nuclear energy for use in weaponry until all the nations of the world were at peace. Rutherford died in 1937, well before the destructive power of atomic energy was unleashed in 1945.

Physics
Greek
phusis- nature
-ic (ikos) relating to or having some characteristic of
The science of matter and energy and of the interactions between the two, grouped into traditional fields such as acoustics, optics, mechanics, thermodynamics, and electromagnetism, as well as modern extensions including atomic and nuclear physics, cryogenics, solid-state physics, particle physics, and plasma physics.

Physiology
Greek
physio- form, origin
-logy (logos) used in the names of sciences or bodies of knowledge
The branch of biology dealing with the structure and functions of living organisms and their parts.

Phytobenthos
Greek
phuton- plant
-benthos deep; the fauna and flora of the bottom of the sea
The aquatic flora of the region at or near the bottom of the sea.

Phytochrome
Greek
phuton- plant
-chrome pigment
A substance that produces a color in plant tissue.

Phytoplankton
Greek
phuton- plant
-planktos wandering
Minute, free-floating aquatic plants.

Pigment
Latin
pingere to paint
A coloring matter in animals and plants, especially in a cell or tissue.

Pineal
French
pomme de pin pinecone
An endocrine gland found in the middle of the brain; it secretes melatonin and is named for its pinecone shape.

Pinniped
Latin
pinnas- feather, wing
-ped foot

Any of a suborder (Pinnipedia) of aquatic carnivorous mammals (such as a seal or walrus) with all four limbs modified into flippers.

Pinocytosis
Greek
pinein- to drink
-kutos- (cyto) sac or bladder that contains fluid
-sis action, process, state, condition
Introduction of fluids into a cell.

Pistil
Latin
pestle club-shaped
The female reproductive organ of a flowering plant; it contains the stigma, style, and ovary.

Pitch
Anglo Norman
piche pitch
The auditory effect of sound frequency; the sap that gathers from evergreen trees; any of the resinous materials from the bitumens, such as asphalt.

Pituitary
Greek
pituitarius- of phlegm
ptuo- to spit
-ary of, relating to, or connected with
A small oval endocrine gland attached to the base of the vertebrate brain, the secretions of which control the other endocrine glands and influence growth, metabolism, and maturation.

Placenta
Greek
plakoenta flat land, surface
A flat, membranous, highly vascular organ that develops in the female mammal during pregnancy; it supplies nutrients and removes wastes from the developing fetus.

Planet
Greek
planasthai to wonder
A heavenly body seeming to have a motion of its own among the fixed stars.

Plankton
Greek
planktos wandering
The passively floating or weakly swimming, usually minute animal and plant life in a body of water.

Plasma
Greek
plastos (plassein) something molded (to mold)
Straw-colored fluid part of the lymph and blood composed of water, electrolytes, proteins, glucose, fats, and gases. Essential for carrying cellular elements of the blood and maintaining acid-base balance.

Plasmalemma
Greek
plastos- (plassein) something molded (to mold)
-eilema veil, sheath
The thin membrane immediately surrounding the cytoplasm of a cell that restricts the passage of molecules into the cell.

Plasmodesmata
Greek
plastos- (plassein) something molded; to mold
-desma bond, adhesion
A strand of cytoplasm that passes through an opening in the cell walls and connects the protoplasts of adjacent living plant cells.

Plasmolysis
Greek
plastos- (plassein) something molded (to mold)
-ly- (luein) to loosen, dissolve, dissolution, break
-sis action, process, state, condition
Contraction of a cell caused by loss of water.

Platyhelminthes
Greek
platus- flat
-helminth worm
Any of various parasitic and nonparasitic worms of the phylum Platyhelminthes, such as a tapeworm or a planarian, characteristically having a soft, flat, bilaterally symmetrical body and no body cavity.

Platypus
Latin
platus- flat
-pous foot
A flat-tailed, semiaquatic mammal, resembling a duck and having webbed feet and a snout; egg laying.

Pleiades
Greek
peleiades flock of doves
The cluster of seven stars also known as the Seven Sisters, located in the constellation Taurus the Bull.

Pleistocene
Greek
pleistos- most
-kainos recent, new
An epoch of the Quaternary period, between 1.8 million years ago and the beginning of the Holocene epoch.

Pleomorphic
Greek
ple- many, more
-morph- shape, form, figure, or appearance
-ic (ikos) relating to or having some characteristic of
Refers to the occurrence of two or more structural forms during a lifespan.

Pleura
Greek
pleura rib, side
Thin membrane that covers a lung and lines the chest cavity in mammals.

Plexus
Greek
plectere to plait, braid
In biology, a network-like structure formed by nerves, blood vessels, or lymphatic vessels.

Pliocene
Greek
pleion- more
-kainos recent, new
Final epoch of the Tertiary period, spanning the time between 5.3 and 1.8 million years ago.

Plutonic
Greek
pluto- the god of the lower world in classical mythology
-ic (ikos) relating to or having some characteristic of
Refers to intrusive rocks that form under the earth's surface.

Pneumonia
Greek
pneumon- lung, breath
-ia names of diseases, place names, or Latinizing plurals
An acute or chronic disease marked by inflammation of the lungs; caused by viruses, bacteria, or other microorganisms and sometimes by physical and chemical irritants.

Pneumonocentesis
Greek
pneumon- lung, breath
-kentesis- pricking
-sis action, process, state, condition
Surgical perforation or puncture of a lung to remove fluid, pus, or blood.

Poikilotherm
Greek
poik- varied
-thermos combining form of "hot" (heat)
An animal that can fluctuate its temperature.

Polar
Greek
polos either of two oppositely charged terminals; axis, sky
Relating to or characterized by a dipole.

Polarity
Greek
polos- either of two oppositely charged terminals; axis, sky
-ity state of, quality of
Intrinsic polar orientation; having two opposite attributes.

Polarization
Greek
polos- either of two oppositely charged terminals, axis, sky
-ar- relating to or resembling
-ize- to cause
-ation act or process
The partial or complete polar separation of positive and negative charges in a nuclear, atomic, or chemical system.

Pollen
Latin
pollen fine flour
Tiny, grainlike structures containing the sperm cells of an angiosperm; they are produced by the anthers of flowers.

Pollination
Latin
pollen- fine flour
-ation act or process
The transfer of pollen to the female cone in conifers or to the stigma in angiosperms.

Polyatomic
Latin
poly- many or much
-atomos- indivisible
-ic (ikos) relating to or having some characteristic of
Consisting of many atoms.

Polycythemia
Latin/Greek
poly- many or much
-cyte- (kutos) sac or bladder that contains fluid
-haima blood
A condition marked by an abnormally large number of red blood cells in the circulatory system.

Polygenic
Greek
poly- many or much
-gen- to give birth, kind, produce
-ic (ikos) relating to or having some characteristic of

Of or relating to more than one gene.

Polyhalophilic
Greek
poly- many or much
-hal- salt
-phile- one who loves or has a strong affinity or preference for
-ic (ikos) relating to or having some characteristic of
Describes marine organisms that thrive in a wide range of salinities.

Polyhedron
Greek
poly- many or much
-hedron head
A three-dimensional, symmetrical shape made up of many faces.

Polyhybrid
Greek
poly- many or much
-hybrida offspring of mixed parents
In genetics, the offspring of parents differing in more than three specific gene pairs.

Polymer
Greek
poly- many or much
-meros a part
A large molecule assembled from small, individual molecules.

Polymerase
Greek
poly- many or much
-meros- parts
-ase enzyme
An enzyme used to convert two or more molecules into a polymer.

Polymorphism
Greek
poly- many or much
-morph- shape, form, figure, or appearance
-ism state or condition, quality
The ability to appear in more than one form.

Polymyalgia
Greek
poly- many or much
-myo- muscle
-algia pain, sense of pain; painful, hurting
Pain affecting several muscles.

Polyp
Greek
poly- many or much
-pous foot

A hydra or coral, having a cylindrical body with a single opening; a nonmalignant tumor or growth extending from the mucosa into the lumen of an organ, such as in the large intestine.

Polypathia
Greek
poly- many or much
-pathos- suffering from
-ia names of diseases, place names, or Latinizing plurals
The presence of several diseases at once.

Polyploidy
Greek
poly- many or much
-ploid- having a number of chromosomes that has a specified relationship to the basic number of chromosomes
-y place for an activity; condition, state
Having one or more extra sets of chromosomes.

Polyprotic
Greek
poly- many or much
-pro-, prot- before, forward; for, in favor of; in front of
-ic (ikos) relating to or having some characteristic of
Of or relating to an acid that can donate more than one proton to a base, or relating to a base that can accept more than one proton.

Polysyndactyl
Greek
poly- many or much
-daktulos toe, finger, digit
Having two or more instances in the same individual of side-to-side fusion of digits.

Polytene
Greek
poly- many or much
-tainia ribbon, tapeworm
Relating to or having large multistranded chromosomes whose corresponding chromomeres are in contact.

Polythetic
Greek
poly- many or much
-thetos- placed
-ic (ikos) relating to or having some characteristic of
Pertains to a category or class that is defined in terms of a broad set of criteria that are neither necessary nor sufficient. Each member of the category must possess a certain minimal number of defining characteristics, but none of the features must necessarily be found in each member of the category.

Polyuria
Greek
poly- many or much
-urea urine
Excessive excretion of urine because of a disease such as diabetes.

Pons
Latin
pons bridge
A bundle of nervous tissue located on the ventral surface of the spinal cord at the base of the brain; it connects the medulla oblongata to higher regions in the brain.

Population
Latin
populus- the people
-ion state, process, or quality of
A group of organisms of the same species living in the same area at the same time.

Porcine
Latin
porc- pig or hog
-ine of or relating to
Of or consisting of swine; related to or resembling swine (pigs and hogs).

Porifera
Latin
porus- pore
-ferre to bear
A pore-bearing organism.

Positron
Greek
posi- positive charge
-tron a particle
The particle having the same mass and spin as an electron but having a +1 charge caused by the interaction of cosmic rays with matter.

Posterior
Latin
post- after, behind
-or a condition or property of things or persons, person who does something
Located behind a part or toward the rear of a structure.

Potential
Latin
poten- power, strength, ability
-ial relating to or characterized by
Describes the energy that an object possesses but has not yet used because of its position or condition.

Pound
Latin
pondo by weight
A unit of weight equal to 16 ounces.

Power
Latin
potis able, powerful
The amount of energy consumed per unit of time.

Precession
Latin
prae- earlier, before, prior to
-cedere- to go
-ion state, process, or quality of
The term used to denote a globe spinning on its axis and describing the wobble as the globe slows down.

Precipitate
Latin
prae- earlier, before, prior to
-capit- to throw headlong, the head
-ate of or having to do with
To cause a solid substance to be separated from a solution.

Precipitation
Latin
prae- earlier, before, prior to
-capit- to throw headlong, the head
-ion state, process, or quality of
Water droplets or ice particles condensed from atmospheric water vapor.

Precocial
Latin
prae- earlier, before, prior to
-coquere- to cook, ripen
-al of the kind of, pertaining to, having the form or character of
Refers to a chick that leaves the nest immediately after hatching.

Predator
Latin
praedari- to prey upon
-or condition or activity
A predatory person, animal, or thing thing that preys upon, devours, or destroys another.

Prehensile
Latin
prehensus to clasp or seize
Refers to appendages that are adapted for clasping or grasping.

Prenatal
Latin
prae- earlier, before, prior to
-nasci be born

-al of the kind of, pertaining to, having the form or character of
Existing or occurring before birth.

Pressure
Latin
premere- to exert steady weight or force against; bear down on
-ura act; process; condition
Force applied uniformly over a surface, measured as force per unit of area.

Prey
Latin
praeda booty, prey
An animal taken by a predator as food.

Primary
Medieval Latin
primus- leader
-ary of, relating to, or connected with
In geology, the term used to describe the characteristics of any rock at the time of its formation. In chemistry, relating to the replacement of one or more atoms by other atoms in a chemical reaction.

Primate
Medieval Latin
primus- leader
-ate characterized by having
A member of the order of mammals that includes monkeys, apes, and humans.

Prism
Greek
prizein to saw off
A piece of glass that is usually cut into a triangular shape so that light can travel through, and so that the colors of the visible light are separated.

Probability
Latin
pro- before; forward; for, in front of; in place of
-abilis- to do something, specific action
-ity state of, quality of
The chance that a given event will occur; a logical relation between statements such that evidence confirming one confirms the other to some degree.

Probiotics
Latin/Greek
pro- before; forward; for, in front of; in place of
-bios- life, living organisms or tissue
-ic (ikos) relating to or having some characteristic of
Beneficial bacteria used to ease digestive ailments.

Proboscidea
Greek
pro- before; forward; for, in front of; in place of
-boskein to feed
Mammalian order that includes elephants.

Prodromal
Greek
pro- before; forward; for, in front of; in place of
-dromos- race course, running
-al of the kind of, pertaining to, having the form or character of
Refers to the time following incubation period when the first signs of illness appear.

Producer
Latin
pro- before; forward; for, in front of; in place of
-duct- lead, take, bring
-er one that performs an action
An organism that has the capacity to make its own food either by photosynthesis or by chemosynthesis.

Product
Latin
pro- before; forward; for, in front of; in place of
-duct lead, take, bring
That which results from the operation of a cause; a consequence, effect.

Prognathous
Greek
pro- before; forward; for, in front of; in place of
-gnathos jaw
Having the head horizontal and the mouthparts directed anteriorly.

Prognosis
Greek
pro- before; forward; for, in front of; in place of
-gnos- know, learn, discern
-sis action, process, state, condition
A prediction of the probable course and outcome of a disease.

Program
Greek
pro- before; forward; for, in front of; in place of
-gramma something written or drawn; a record
Data instructions fed into a computer to control the actions of the computer.

Prokaryotic
Greek
pro- before; forward; for, in front of; in place of
-karyon- kernel, nucleus
-ic (ikos) relating to or having some characteristic of
Lacking a membrane-bound nucleus and membranous organelles, as in bacteria and archaea.

Prominence
Greek/Latin
pro- before; forward; for, in front of; in place of
-minere- to jut or threaten
-ence the condition of
The incredibly huge masses of gases that burst forth from the chromosphere of the sun.

Pronotum
Greek
pro- before; forward; for, in front of; in place of
-noton- the back
-um (**singular**) structure
-a (**plural**) structure
The upper, often shieldlike, hardened body-wall plate located just behind the head of an insect.

Propagation
Latin
pro- before; forward; for, in front of; in place of
-pangere- to fasten
-ate- of or having to do with
-ion state, process, or quality of
The multiplication or natural increase in a population; the dissemination of something to a larger area or greater number.

Propellent
Latin
pro- before; forward; for, in front of; in place of
-pellere- to drive
-ant a person who, the thing which
The fuel and oxidizer of a rocket that provides the thrust needed for the rocket to escape earth's gravity.

Prophase
Greek
pro- before; forward; for, in front of; in place of
-phainein to show
The stage of cell division in which the chromosomes condense and become visible.

Prosencephalon
Greek
pro- before; forward; for, in front of; in place of
-enkephalos in the head
The anterior portion of the forebrain, including the frontal lobe and the olfactory bulbs.

Prosimians
Latin
pro- before; forward; for, in front of; in place of
-simia- ape, monkey
-an one that is of or relating to or belonging to
Of or belonging to Prosimii, a suborder of primates that includes the lemurs, lorises, and tarsiers.

Prostate
Greek

pro- before; forward; for, in front of; in place of
-histanai to set, place
A gland that wraps around the urethra in males. It is responsible for releasing urine from the urinary bladder to the exterior, and it produces seminal fluid, a principal component of semen.

Protactinium
English
pro-, prot- before, forward; for, in favor of; in front of
-actinium element actinium
A rare, extremely toxic radioactive element, which decays into actinium.

Protandrous
Greek
pro-, prot- before, forward; for, in favor of; in front of
-andr- man, male, men, masculine
-us thing
Of or relating to a flower in which the anthers release their pollen before the stigma of the same flower is receptive.

Protection
Latin
pro-, prot- before, forward; for, in favor of; in front of
-tegere- to cover, ward off, guard, defend
-ion state, process, or quality of
The act of safeguarding, preserving, or shielding.

Protective
Latin
pro-, prot- before, forward; for, in favor of; in front of
-tegere- to cover, ward off, guard, defend
-ive performing an action
Describes the act of guarding another person from danger or injury and providing a safe environment.

Protein
French
proteine of the first quality
Any group of complex organic macromolecules containing carbon, hydrogen, oxygen, nitrogen, and usually sulfur. Proteins are composed of one or more chains of amino acids and include many substances, such as enzymes, hormones, and antibodies, that are necessary for the proper functioning of an organism.

Proteolysis
Greek
prote- protein
-ly- (*luein*) to loosen, dissolve; dissolution, break
-sis action, process, state, condition
A reaction sequence of the noncyclic pathway of photosynthesis, triggered by photon energy, in which water is split into oxygen, hydrogen, and electrons.

Proterozoic
Greek
proteros- earlier
-zoikos- of animals
-ic (ikos) relating to or having ome characteristic of
Relating to the geologic era characterized by the first signs of single-celled organisms, plant algae.

Protist
Latin
protos- first formed, original, earliest
-ist performs an action
Unicellular organism belonging to kingdom Protista.

Protium
Greek
protos- first formed, original, earliest
-ium chemical element
The most abundant isotope of hydrogen, with atomic mass of 1.

Protocell
Greek/Latin
protos- first formed, original, earliest
-cella chamber
A structure that has a lipid protein membrane and carries on energy metabolism it existed before the first true cell.

Protogynous
Greek
protos- first formed, original, earliest
-gune woman, women, female
Referring to animals that are sequential hermaphrodites, where that animal is first biologically female, having only female sexual organs, and then changes to become biologically male.

Protolithic
Greek
protos- first formed, original, earliest
*-lith- r*ock, stone
-ic (ikos) relating to or having some characteristic of
Of, relating to, or characteristic of the very beginning of the Stone Age; Eolithic.

Proton
Greek
protos- first formed, original, earliest
-on a particle
An elementary particle that is identical to the nucleus of the hydrogen atom, that along with neutrons is a constituent of all other atomic nuclei, that carries a positive charge numerically equal to the charge of an electron, and that has a mass of $1.673 \, Þ \, 10^{-27}$ kg.

Protoplast
Greek
protos- first formed, original, earliest
-plastos (plassein) something molded (to mold)
Plant cell from which the cell wall has been removed.

Protostome
Greek
protos- first formed, original, earliest
-stoma mouth
An animal whose mouth develops from or near the blastopore; an opening in the early embryo.

Prototheria
Greek
protos- first formed, original, earliest
-theria wild animal, monotremes
Subclass of Cretaceous and early Cenozoic mammals; extinct except for egg-laying monotremes.

Prototype
Greek
protos- first formed, original, earliest
-tupos impression
An original type, form, or instance serving as a basis or standard for later stages.

Protozoa
Greek
protos- first formed, original, earliest
-zoan animal, living being; life
Single-celled microorganisms of the sub-kingdom Protozoa; lowest form of animal life.

Proximity
Latin
proximus- nearest, next
-ity state of, quality of
The state, quality, sense, or fact of being near or next to; closeness.

Pseudocoelom
Greek
pseudes- false
-koiloma cavity
Body cavity lying between the digestive tract and body wall.

Pseudopodia
Greek
pseudes- false
-podion base, foot
A fingerlike projection on the body of an amoeba used for movement.

Psychokinesis
Greek
psych- mind, consciousness, mental process
-kinetikos- to move; set in motion
-sis action, process, state, condition
The production or control of motion by a subject without any intermediate physical energy.

Psychosomatic
Greek
psych- mind, consciousness, mental process
-soma- (somatiko) body
-ic (ikos) relating to or having some characteristic of
Of or relating to a disorder having physical symptoms but originating from mental or emotional causes.

Psychrometer
Greek
psychros- cold
-meter (metron) instrument or means of measuring; to measure
Instrument that measures humidity.

Pterodactyl
Greek
pteron- feather, wing
-daktulos toe, finger, digit
Small, typically tail-less winged reptile existing in the Jurassic and Cretaceous periods.

Pterygoid
Greek
pterug- wing
-oid (oeides) resembling, having the appearance of
Relating to the region of the sphenoid bone of the skull; winglike muscle.

Pulmonary
Latin
pulmo- lung
-ary of, relating to, or connected with
Relating to or involving the lung.

Pulsar
Latin
pullere- to beat
-ar relating to or resembling
A relatively small star composed of neutrons that emit radiant energy in regular pulses.

Pupil
Latin
pupilla little doll; pupil of the eye (named for the tiny reflections on the eye)
The hole in the center of the iris that light travels through in order to be focused on the retina.

Purine
Latin
purus- clean
-ine of or relating to
The nitrogenous bases, adenine and guanine, found in DNA.

Putrefaction
Latin
putrefacere- to make rotten
-ion state, process, or quality of
The process of creating a strong, foul odor by emitting gases from the decomposition of organic material.

Pylorus
Latin
pule- gate
-ouros guard
The lower section of the stomach that includes the passageway into the duodenum of the small intestine.

Pyrimidine
Latin
pur- fire
-ide- group of related chemicals
-ine of or relating to
The nitrogenous bases, cytosine and thymine, found in DNA.

Pyroclastic
Greek
pur- fire
-klastos broken
Composed chiefly of rock fragments of volcanic origin.

Pyroxenes
Greek
pur- fire
-xenos stranger
Any of a group of crystalline silicate minerals common in igneous and metamorphic rocks and containing two metallic oxides.

Pyrrole
Greek
pyre- red
-ole a heterocyclic chemical with a five-membered ring
A five-membered heterocyclic ring compound, C_4H_5N, that has an odor similar to chloroform and is the parent compound of hemoglobin.

Pyuria
Greek
puo- pus
-uria urine
Pus found in the urine; usually an indication of an infection.

Q

Quadriceps
Latin
quadi- four
-caput head
A very large muscle on the anterior surface of the thigh; it contains four heads (cusps).

Quadruped
Latin
quadi- four
-ped foot
A four-footed animal that uses all four feet for walking and running.

Quantum
Latin
quantus how great
The smallest amount of a physical quantity that can exist independently, especially a discrete quantity of electromagnetic radiation.

Quartz
German
quarz mineral quartz
A very hard mineral composed of silica.

Quasar
English

quasi- having a likeness to something
-(stell)ar star
A starlike object that has a large red shift and emits powerful blue light and often radio waves.

Quaternary
Latin
quartern- four
-ary of, relating to, or connected with
The second period of the Cenozoic era, spanning the time between 1.8 million years ago and the present.

Quiescence
Latin
quies- still, quiet
-ence the condition of
A state in which a seed or other plant will not germinate or grow until the requisite environmental conditions occur.

Quintessence
Latin
quinta- fifth
-essentia essence
The fifth or last and highest essence in ancient and medieval philosophy, above fire, air, water, and earth, that permeates all nature and is the substance composing the heavenly bodies.

R

Rabies
Latin
rabere to rave
A fatal disease caused by a virus that is transmitted by a mammal; the symptoms include hydrophobia, convulsions, heightened excitability, and muscular spasms in the throat.

Radial
Latin
ray- spoke of a wheel
-ial relating to or characterized by
Of or characterized as being arranged in a raylike fashion.

Radiant
Latin
radiare to radiate
Of or referring to energy traveling by means of electromagnetic waves.

Radioactivity
English
radi- radiant or radiation energy; wireless transmitter
-agere- drive, do
-ity state of, quality of
The emission of radiation, either spontaneously from unstable atomic nuclei or as a consequence of a nuclear reaction.

Radionuclide
English/Latin
radi- radiant or radiation energy; wireless transmitter
-nucula- kernel, little nut
-ide nonmetal radical
A radioisotope; a nuclide that exhibits a certain amount of radioactivity.

Radiosonde
English/French
radi- radiant or radiation energy; wireless transmitter
-sonde a sounding lead/line
A measurement device that is carried aloft by a balloon to relay temperature, pressure, and humidity data from the upper atmosphere.

Radius
Latin
ray- spoke of a wheel
-ius singular
A line segment that connects the center of a circle or sphere to any point on its outer edge.

Radula
Latin
radere to scrape
Flexible, tonguelike organ in certain mollusks, having rows of horny teeth on the surface.

Range
German
reng to put in a row, line
The difference between the smallest and largest values in a distribution.

Raptor
Latin
rapere to seize
A bird of prey; carnivorous bird that hunts its prey.

Joseph Meister Had Rabies

On a sunny day in the summer of 1885 at Meissen-gott, in Alsace, a boy named Joseph Meister was attacked by a neighborhood dog. The 9-year-old Joseph was thrown to the ground, and as he tried to protect his face he was savagely bitten about the arms. The dog was finally driven off the boy, but the damage was done. His skin had been pierced by a rabid dog.

The local physician did all he could. He cauterized and cleaned the wounds, but he knew what would soon happen to the child. He advised the mother to take him to Louis Pasteur, a scientist who was experimenting on rabies in Paris. Though Pasteur was not a physician, he was the boy's best and only hope.

Once bitten by a rabid animal, the human victim experiences a brief period of fever and restlessness before becoming wildly excitable. The infected individual salivates excessively and a white, frothy foam appears around the mouth. The muscles of the throat become highly irritated, with uncontrollable spasms causing great pain. All the while the victim experiences an uncontrollable thirst for water but is unable to drink. This torture continues relentlessly for up to five days before the victim falls dead as a result of exhaustion, asphyxia, and paralysis.

What could cause such horrible symptoms? Rabies was a disease known to the ancients. Although it was never the blight that the plagues that ravaged Europe and Asia were, it brought fear to those who witnessed the agonizing death of its victims. The Greeks attributed rabies to the wrath of the gods. Sirius, the Dog Star, in the constellation Canis Major, was believed to be the cause of the disease. The days during summer in the Northern Hemisphere when Sirius rises immediately before or sets immediately after the sun, referred to as the "dog days," were believed to be a time when normally docile animals would run wild and become viciously aggressive. Shortly thereafter, they would convulse, become paralyzed, and die.

In the fifth century BC, the Greek physician Democritus described the symptoms of rabies, as did Aristotle two hundred years later. The Romans in the first century AD cauterized or placed the ashes of seahorses on the wounds to treat the condition, but, of course, these treatments were futile.

Pliny the Elder, a Roman naturalist who lived in the first century AD, wrote on the treatment of rabies:

It is universally agreed, too, that when a person has been bitten by a dog and manifests a dread of water and of all kinds of drink, it will be sufficient to put under his cup a strip of cloth that has been dipped in menstrual fluid; the result being that the hydrophobia will immediately disappear. This arises, no doubt, from that powerful sympathy which has been so much spoken of by the Greeks, and the existence of which is proved by the fact, already mentioned, that dogs become mad upon tasting this fluid.

When Joseph Meister and his mother arrived in Paris on July 6, Joseph was in very bad shape. His pain was such that he could barely walk. Pasteur knew what he had to do, but he needed to consult with colleagues. According to Pasteur, the numerous trials of his rabies vaccine on animals had proven to be a resounding success. Later we would find out otherwise, but nonetheless, this was Joseph's last and best chance at survival. A team of government scientists gave Pasteur their approval to begin the procedure.

Over the next 11 days, Joseph was injected with small amounts of the vaccine, which Pasteur had prepared using the spinal cords of infected rabbits. Pasteur wrote in his journal:

The death of this child appearing to be inevitable, I decided, not without lively and sore anxiety, as may well be believed, to try upon Joseph Meister, the method which I had found constantly successful with dogs. Consequently, sixty hours after the bites, and in the presence of Drs Vulpian and Grancher, young Meister was inoculated under a fold of skin with half a syringeful of the spinal cord of a rabbit, which had died of rabies. It had been preserved (for) fifteen days in a flask of dry air. In the following days, fresh inoculations were made. I thus made thirteen inoculations. On the last days, I inoculated Joseph Meister with the most virulent virus of rabies.

There were side effects—Joseph experienced bouts of anxiety and depression—but there were no longer signs of the dreaded disease. And so, after ten more days of observation, Joseph was sent home. He had escaped death.

Years later, Joseph Meister would return to Paris and work as doorman for the Pasteur Institute. He worked at the institute until the age of 64 in 1940, when the Nazis invaded Paris. The Germans ordered Meister to open Pasteur's crypt. Rather than obey that order, Joseph Meister put a gun to his head and ended his own life.

Marie Curie

Eve Curie wrote of her mother, "She was a woman; she belonged to an oppressed nation; she was poor; she was beautiful. A powerful vocation summoned her from her motherland, Poland, to study in Paris, where she lived through years of poverty and solitude. There she met a man. . . . By the most desperate and avid effort they discovered a magic element, radium. This discovery not only gave birth to a new science and new philosophy; it provided mankind with the means of treating a dreadful disease."

Marie Curie was born Marie Sklodovska in Poland on November 7, 1867. She had a rather distress-filled youth. Her sister died of typhus and her mother passed away four years later. After her high school years Marie sunk into a depressive state.

Marie showed signs of brilliance at a young age. She possessed an amazing memory and an intellectual curiosity, but attending a university in Poland was out of the question. She knew that to thrive, she would have to leave Poland. Years later, in Paris, after studying physics and chemistry at the University of Paris (Sorbonne), she became the first woman to teach at that highly prestigious institu-tion. There she met Pierre Curie, whose title was Chief of the Laboratory of the School of Physics and Chemistry of the City of Paris. They married and together studied radiation and subsequently discovered the elements radium and polonium.

Her work led to the use of x-rays in World War I. This remarkable application of radiation allowed surgeons to more easily find the bullets lodged in soldiers, giving them a greater chance of survival through surgery. Her studies with radiation led to additional research on the role of radiant energy in the reduction of cancerous growths. Her accomplishments led her to become the first person to receive Nobel Prizes in two different fields of study, physics and chemistry. This feat has been matched only by Linus Pauling, who won Nobel Prizes for Chemistry and Peace.

Ironically, her isolation of the radioactive materials from the ore pitchblende for the advancement of science and medicine ultimately led to her own death from leukemia in 1934. Albert Einstein said of Madam Curie, "Marie Curie is, of all celebrated beings, the only one whom fame has not corrupted."

Rarefaction
Latin
rarus- rare
-facere- to make
-ion state, process, or quality of
That part of the sound wave where the particles of the sound medium are farthest apart.

Rate
Latin
rata according to a fixed proportion
A quantity, amount, or degree of something measured per unit of time.

Ratiocination
Latin
ratio- reason
-cinari- reckon
-ion state, process, or quality of
To reason using formal logic; to use deductive reasoning.

Rawinsonde
English/French
radi- radiant or radiation energy; wireless transmitter
-wind- moving air
-sonde a sounding lead/line
A radiosonde used to observe the velocity and direction of upper-air winds and tracked by a radio direction-finding instrument.

Reactance
Latin
re- to do something again; go against
-agere to drive, do
Opposition to the flow of alternating current caused by the inductance and capacitance in a circuit rather than by resistance.

Reaction
Latin
re- to do something again; go against
-agec- to act
-ion state, process, or quality of
A response in opposition to a substance, treatment, or other stimulus.

Reactive
English/Latin
re- to do something again; go against
-agec- to act
-ive performing an action
Tending to participate readily in reactions.

Reagent
English/Latin
re- to do something again; go against
-agere a force or substance that causes a change
A substance used in a chemical reaction to detect, analyze, or produce other substances.

Receptor

Latin

reciepere to receive

A group of sensory nerve endings that respond to threshold energy from a source point.

Recessive

Latin

recedere- to recede

-ive performing an action

In genetics, refers to an allele that does not display its phenotype when paired with a dominant gene.

Reclamation

English/Latin

re- to do something again; go against

-clamare- to call or cry out

-ion state, process, or quality of

The act or process of reclaiming; restoration for the purpose of productivity.

Rectifier

Latin

rectus- straight, direct

-er one that performs an action

A device, such as a diode, that converts alternating current to direct current.

Rectoclysis

Latin

rectus- straight, direct

-clys, -clysis to wash, washing

Washing or irrigation of the rectum.

Recycle

English/Greek

re- to do something again; go against

-kyklos circle, wheel, cycle, rotate

To make ready for reuse; to pass again through a series of changes or treatments.

Reduction

English/Latin

re- to do something again; go against

-ducere- to lead

-ion state, process, or quality of

To decrease the valence of an atom by adding electrons.

Reflectivity

English/Latin

re- to do something again; go against

-flectere- to throw or bend back

-ity state of, quality of

The ratio of the energy of a wave reflected from a surface to the energy possessed by the wave striking the surface.

Reflux

Latin

re- to do something again; go against

-fluere to flow, wave

A flowing back, ebb; the process by which a container with boiling liquid is attached to an apparatus that continuously returns the vapor for reboiling.

Reform

English/Latin

re- to do something again; go against

-forma shape, figure, appearance

To improve by alteration, correction of error, or removal of defects; put into a better form or condition.

Refraction

English/Latin

re- to do something again; go against

-fract- to break

-ion state, quality, or process of

The turning or bending of any wave, such as a light or sound wave, when it passes from one medium into another of different optical density.

Regolith

Greek

rhegos- blanket

-lith rock, stone

The layer of loose rock resting on bedrock, constituting the surface of most land.

Relay

English/French

re- to do something again; go against

-laier to leave

An electrical device used to control a switch or to allow a weak current to control a stronger electrical current.

Relief

French

relever to relieve

The difference in height from the lowest to the highest point.

Renal

Latin

reno- kidney

-al of the kind of, pertaining to, having the form or character of

Of or relating to the region of the kidneys.

Reniform

Latin

renes- kidney

-forma having the form of

Being in the shape of a kidney, such as a leaf.

Replicase

English/Latin
re- to do something again; go against
-plicare- to fold
-ase enzyme
An enzyme that catalyzes the synthesis of a complementary RNA molecule from an RNA template.

Replicate

English/Latin
re- to do something again; go against
-plicare- to fold
-ate characterized by having
To reproduce or make an exact copy or copies of genetic material.

Repressor

Latin
re- to do something again; go against
-premere- to press back
-or a condition or property of things or persons; person who does something
A protein produced by the regulator gene; it blocks the transcription of the gene.

Reproduction

English/Latin
re- to do something again; go against
-pro- before; forward; for, in front of; in place of
-ducere- to lead
-ion state, process, or quality of
The act of (re)producing something of the same kind.

Reside

Latin
residere to sit back, abide, remain
To dwell permanently or continuously.

Resistance

English/Latin
re- to do something again; go against
-sistere- to place
-ance brilliance, appearance
A force that tends to oppose or retard motion.

Resistor

English/Latin
re- to do something again; go against
-sistere- to place
-or a condition or property of things or persons; person who does something
A component that resists the flow of current in an electronic circuit.

Resolution

Latin
resolvere- relax, untie
-ion state, process, or quality of
The process of distinguishing the individual parts of an object.

Resonance

Latin
re- to do something again; go against
-sonare- to sound
-ant performing, promoting, or causing a specified action
The condition that causes a medium to vibrate in its natural frequency as a result of receiving sound waves of the same frequency.

Respiration

English/Latin
re- to do something again; go against
-spire- to breathe
-ion state, process, or quality of
The molecular exchange of oxygen and carbon dioxide within the body's tissues, from the lungs to the cellular oxidation processes; the act of inhaling and exhaling.

Response

Latin
re- to do something again; go against
-spondere to promise
The reaction by a living organism to a stimulus.

Restitution

English/Latin
re- to do something again; go against
-statuere- to set up
-ion state, process, or quality of
The return to or restoration of a previous state or position after a collision.

Resultant

English/Latin
re- to do something again; go against
-saltare to leap
A vector generated through the sum of other vectors.

Retardant

Latin
re- to do something again; go against
-tardare- delay, impede
-ant performing, promoting, or causing a specified action
Acting or intending to delay or impede. This term is often used with another term, as in "flame retardant."

Reticulum

Greek/Latin
reticul- net or networklike
-um (**singular**) structure
-a (**plural**) structure
System of membranous saccules and channels in the cytoplasm, often with attached ribosomes.

Retina
Latin
retis net
Innermost layer of the eyeball.

Retrovirus
Latin
retro- backward, behind
-virus poison
A group of viruses each of which contains one strand of RNA. The group includes many viruses that may cause some cancers, as well as the HIV virus.

Revolution
Latin
re- to do something again; go against
-volvere- to turn or spin
-ion state, process, or quality of
The movement of one body (planet) around another body (sun) or a fixed point.

Rex
Latin
rex king
The king; refers to or denotes size or dominance of a given species (e.g., *Tyrannosaurus rex*).

Rheumatic
Greek
rheum- flow, watery discharge from the body once thought to cause aches and pains in joints
-ic (ikos) relating to or having some characteristic of
Of, relating to, or having the characteristics of rheumatism.

Rheumatism
Greek
rheumat- flow, watery discharge from the body once thought to cause aches and pains in joints
-ism state or condition, quality
Any of a number of pathological conditions leading to mild to severe aches and pains in the joints.

Rhinencephalon
Greek
rhin- nose
-cephalo- (kephalikos) head
-on a particle
That portion of the cerebrum concerned with reception and integration of olfactory (smelling) impulses.

Rhinitis
New Latin
rhin- nose
-itis inflammation, burning sensation
Inflammation of the mucous membranes of the nose.

Rhinoceros
Latin
rhin- nose
-keras horn
Any of a family (Rhinocerotidae) of large, heavy-set, herbivorous perissodactyl mammals of Africa and Asia that have one or two upright keratinous horns on the snout and thick gray to brown skin with little hair.

Rhinomycosis
Greek
rhin- nose
-myco- (mukes) fungi
-sis action, process, state, condition
Fungal infection of the nasal mucous membranes.

Rhinorrhea
New Latin
rhin- nose
-rhea flow or discharge
Secretions or discharge from the nose.

Rhizobium
Greek
rhiza- root
-bios- life, living organisms or tissue
-um (**singular**) structure
-a (**plural**) structure
Any of various nitrogen-fixing bacteria of the genus *Rhizobium* that form nodules on the roots of leguminous plants, such as clover and beans.

Rhizoid
Greek
rhiza- root
-oid (oeides) resembling, having the appearance of
Rootlike hair that anchors a plant and absorbs minerals and water from the soil.

Rhodophyte
Greek
rhodon- rose
-phyte plant
Marine algae with a reddish color or hue.

Ribonucleic acid
German/Latin
ribo(se)- a kind of sugar
-nucula- kernel, little nut
-ic (ikos) relating to or having some characteristic of
A long, single-stranded polymer found in all living organisms and involved in genetic transcription and protein synthesis.

Ribosome
Greek
ribose- sugar
-soma (somatiko) body
A minute, round particle composed of RNA and protein, found in the cytoplasm of living cells and active in the synthesis of proteins.

Rigid

Latin

rigere to be stiff

Refers to a system of particles whose positions remain fixed relative to each other.

Riparian

Latin

ripa- river bank, stream

-an one that is of, or relating to, or belonging to

Relating to or living on or near the banks of a stream or river.

Robot

Czech

robot worker

A machine in the form of a human being that performs the mechanical functions of a human being but lacks emotions and sensitivity.

Rodent

Latin

rodere to gnaw

Any member of the order Rodentia, a group of animals in the class Mammalia characterized by having fur, four legs, warm blood, and large incisors for gnawing.

Rodenticide

Latin

rodere- to gnaw

-cide (caedere) to cut, kill, hack at, or strike

A type of pesticide that controls mice, rats, and other rodents.

Rostrum

Latin

rostrum beak

A beaklike or snoutlike projection.

Rotation

Latin

rota- wheel

-ion state, process, or quality of

The act or process of turning about a center or an axis.

S

Saccharide
Sanskrit
sarkara- sugar
-ide group of related chemical compounds
Another name for a sugar.

Saccharolytic
Sanskrit/Greek
sarkara- sugar
-ly- (luein) to loosen, dissolve, dissolution, break
-ic (ikos) relating to or having some characteristic of
Capable of hydrolyzing or otherwise breaking down a sugar molecule.

Sacrum
Latin
sacr- sacred or holy
-um (**singular**) structure
-a (**plural**) structure
Compound triangular bone at the base of the human spine.

Sagittal
Latin
sagitta- arrow
-al of the kind of, pertaining to, having the form or character of
Relates to the plane that is parallel to the sagittal suture of the skull.

Salamander
Latin
salamandra slithering
Any member of the order Caudata, having porous, smooth skin, weak legs, and a tail.

Salt
Old English
sealt salt
A compound created by the neutralization of an acid with a base or by a chemical reaction between a metal and a nonmetal.

Saponification
Latin
saponi- soap
-fication to make
The process of saponifying; the decomposition of a fat by the addition of an alkali that combines with its fatty acids to form a soap, with the remaining constituent, glycerin, consequently liberated.

Saprophagous
Greek
sapro- rotten, putrid; decay
-phagos (phagein) to eat, eating
Feeding on decaying matter; carrion beetles who feed off of the rotting matter of dead organisms.

Saprophyte
Greek
sapro- rotten, putrid; decay
-phyton plant
A plant living on dead or decaying organic matter.

Saprotroph
Greek
sapro- rotten, putrid, decay
-trophos (trophein) to nourish, food, nutrition; development
Organism that secretes digestive enzymes and absorbs the resulting nutrients back across the plasma membrane.

Sarcolemma
Greek
sarko- flesh, meat
-eilema veil, sheath
The plasma membrane of a muscle cell.

Sarcoma
Greek
sarko- flesh, meat
-oma tumor
Cancerous tumor derived from connective tissue.

Sarcomere
Greek
sarko- flesh, meat
-mere part, segment
A segment of a striated muscle cell fibril bounded by Z-disks.

Satellite
French/Latin
satelles- to hang on
-ite component of a part of a body
A celestial body (moon) revolving around another celestial body (planet).

Saturated
Latin
satur- full
-ate characterized by having
Incapable of holding any more of a substance or material.

Saurischia
Greek
sauros- lizard
-iskhion hip joint
A dinosaur of the order Surischia characterized by having the pelvic girdle of a modern-day reptile.

Scapula
Latin
scapulae shoulder blade
A triangular bone forming the dorsal part of the shoulder.

Schistosome
Greek
skhizein- to cut, split
-soma (somatiko) body
Any of several chiefly tropical trematodes (worms of the genus *Schistosoma*), many of which are parasitic in the blood of humans and other mammals.

Schizocarp
Greek
skhizein- to cut, split
-karpos fruit
Fruit that splits into several closed, one-seeded portions upon maturation.

Schizocoelus
Greek
skhizein- to cut, split
-koilos hollow
The type of development found in protosomes; initially solid masses of mesoderm split to form coelomic cavities.

Science
Latin
scire to know, knowledge
The observation, identification, description, experimentation, investigation, and theoretical explanation of phenomena.

Scientific
Latin
scire- to know, knowledge
-ic (ikos) relating to or having some characteristic of
Relating to or employing the methodology of science.

Scintillation
Latin
scintilla- spark
-ion state, process, or quality of
A flash of light produced in a phosphor by absorption of an ionizing particle or photon.

Scion
Old French
cion descendant
A grafted twig or bud.

Sclera
Greek
skleros hard
Outer, white, fibrous layer of the eye that surrounds the eye except for the transparent cornea.

Sclerenchyma
Greek
sklero- hard
-en- in
-khein to pour
A supportive plant tissue that consists of thick-walled, usually lignified cells.

Scoliosis
Greek
skolios- crookedness
-osis disease or abnormal condition
Abnormal lateral curvature of the vertebral column.

Scorpio
Greek
skorpios scorpion
The constellation (also called the Scorpion) that lies near Libra and contains the bright red star Antares.

Seamount
Middle English/Latin
see- sea
-mons mountain
A submarine mountain rising more than 500 fathoms (3,000 feet) above the ocean floor.

Secretion
Latin
secernere- to set aside
-ion state, process, or quality of
The state or process of secreting a fluid. Typically these substances are not waste products; they include hormones, mucus, and enzymes.

Sedative
Latin
sedates- to calm
-ive performing an action
A drug that reduces excitability and calms a person.

Sediment
Latin
sed- sit
-ment state or condition resulting from a (specified) action
To sit, sink down; the matter that settles to the bottom of a liquid.

Sedimentation
Latin
sed- sit
-ment- state or condition resulting from a (specified) action
-ation act or process
The act or process of depositing sediment or gravel as a result of some outside force.

Seismograph
Greek
seismos- to shake
-graphia (graphein) to write, record, draw, describe
Instrument used to detect and record seismic waves produced by earthquakes.

Seismologist
Greek
seismos- to shake
-logist a person who studies
A person who studies earthquakes.

Selenium
Greek
selene- moon
-ium quality or relationship
A nonmetallic element resembling sulfur and obtained primarily as a by-product of copper refining; used in photocells.

Semipermeable
Latin
semi- half
-per- through
-meare- to glide
-able capable, be inclined to, tending to, given to
Partially permeable; refers specifically to a membrane that allows smaller objects to pass through while prohibiting larger ones.

Senescence
Latin
sen- old age
-esce- beginning, becoming
-ence the condition of
The sum of processes involving aging, decline, and eventual death.

Sensitivity
Latin/Greek
sensus- sense
-ive- performing an action
-ity state of, quality of
The capacity of an organism to be aware of a stimulus.

Sepsis
Greek
sepein- to make rotten, putrefactive
-sis action, process, state, condition
A poisoned condition resulting from pathological organisms or their toxins in the circulatory system.

Septic
Greek
sepein- to make rotten, putrefactive
-ic (ikos) relating to or having some characteristic of
Relates to the process of living tissue becoming poisoned or rotten as a result of a pathological organism.

Septicemia
Greek
sepein- to make rotten, putrefactive
-haimo- relating to blood or blood vessels
-ia names of diseases, place names, or Latinizing plurals
A systemic disease caused by pathogenic organisms or their toxins in the bloodstream; also called blood poisoning.

Septum
Latin
saepire- to enclose
-um (**singular**) structure
-a (**plural**) structure
A partition or membrane that separates one cavity or hollow from another.

Sessile
Latin
sessus- to sit
-ile changing, ability, suitable, tending to
Without petiole or pedicel—attached directly to the base; fixed, nonmotile animal.

Setae
Latin
seta bristle
Slender, usually rigid or bristly, and springy organ or part of animal or plant.

Sextant
Latin
sextus sixth
An instrument so named because it is a sixth of a circle. It is used to determine latitude and longitude by measuring the altitude of a star or the sun above the horizon.

Shadow zone
Old English
sceadu shade, shadow
The region on the earth's surface ranging from about 7,000 to 10,000 miles from an earthquake in which a seismograph detects no S waves and few, weak P waves.

Sidereal
Latin
sidereus- constellation, star
-al of the kind of, pertaining to, having the form or character of
Of, relating to, or concerned with the stars or constellations; stellar.

Sideropenia
Greek
sideros- iron
-penia reduction, poverty, lack, deficiency
An abnormally low concentration of serum iron in the blood.

Silicate
Latin
silex- hard stone flint
-ate characterized by having
Any of a large group of minerals, forming over 90% of the earth's crust, that consist of SiO_2 or SiO_4 groupings combined with one or more metals and sometimes hydrogen.

Silurian
Celtic
silures- a tribe of Wales
-an one that is of, or relating to, or belonging to
Geologic period in the Paleozoic era that marked the first appearance of air-breathing animals.

Silver
Middle English/Assyrian
siolfor to smelt, refine
sarapu refined silver
The metallic element with atomic number 47, highly valued for its luster.

Simultaneous
Latin
simul- at the same time
-eous having the quality of, relating to
Happening, existing, or done at the same time.

Sinoatrial node
Latin
sinus- hollow
-atri- open area, central court, hall, entrance, or main room of an ancient Roman house
-ium quality or relationship
A small mass of cardiac tissue located in the posterior wall of the right atrium, sometimes referred to as the pacemaker.

Sinus
Latin
sinus hollow
A cavity or depression formed by a series of curved surfaces within a living organism, as in the human skull.

Siphonaptera
Latin/Greek
siphon- siphon
-apteros wingless
Small, wingless, bloodsucking insects with mouthparts adapted for siphoning body fluids from their victims; fleas.

Sirenia
Greek
siren- group of female, partly human creatures in Greek mythology that lured mariners to destruction by their singing
-ia names of diseases, place names, or Latinizing plurals
Herbivorous marine mammals, including the manatee and the dugong.

Skeleton
Greek
skeletos dried up
The bony framework of the body that provides structure, protection, storage of minerals, and an environment for hematopoeisis.

Society
Latin
socius companion, fellowship

An organized population or colony, sometimes having a division of labor.

Sociobiology
Latin
socius- companion or partner
-bios- life, living organisms, or tissue
-logy (logos) used in the names of sciences or bodies of knowledge
The study of the biological basis of all social behavior.

Soil
Latin
solium seat, soil
The top layer of the earth's surface, consisting of rock and mineral particles mixed with organic matter.

Sol
Latin
sol one, alone, or only
Colloid of very small, solid particles dispersed in a liquid that retains the physical properties of a liquid.

Solar
Latin
sol- the sun
-ar relating to or resembling
Of, relating to, or proceeding from the sun.

Solenoid
Greek
solen- pipe
-oid (oeides) resembling, having the appearance of
A coil of wire that acts like a magnet when a current passes through it.

Solid
Latin
solidus firm, unyielding, whole, entire
Matter that has both a definite shape and a definite volume.

Solstice
Latin
sol- the sun
-status to come to a stop, to stand
The two points along the earth's elliptical orbit where the sun's distance from the equator is greatest.

Soluble
Latin
solvere- to loosen
-able/-ible capable, be inclined to, tending to, given to/capable
Describes the ability to be homogeneously mixed in another substance.

Solution
Latin
solvere- to loosen
-ion state, process, or quality of
The process of forming a homogeneous mixture of any combination of solids, liquids, and gases.

Somatic
Greek
soma- (somatiko) body
-ic (ikos) relating to or having some characteristic of
Having to do with the body or body cavities or cells other than reproductive cells.

Somatotropin
Greek
soma- (somatiko) body
-trope- bend, curve, turn, a turning; response to stimulus
-in protein or derived from a protein
Hormone released by the anterior pituitary that stimulates growth in humans.

Somnambulism
Latin
somnia- sleep; dream
-ambulate- walk, take steps, move around
-ism state or condition, quality
Sleepwalking or the ability to perform activities normally associated with being awake while actually sleeping.

Sonoluminescence
Latin
sonus- sound
-lumen- light
-ence the condition of
The production of light as a result of the passing of sound waves through a liquid medium. Light is formed when bubbles in the liquid burst and release energy.

Sorus
Greek
soros a heap
A cluster of sporangia borne on the underside of a fern frond.

Spathe
Latin
spatha a flat blade
A large, leaflike part enclosing a flower cluster.

Speciation
Latin
species- particular kind
-ation state, process, or quality of
Emergence of a new species during evolutionary history.

Species

Latin

species particular kind

A taxonomic unit ranking below a genus and designated by a binomen consisting of its genus name and the species name.

Specimen

Latin

specere to look at, appearance

A small sample of something intended to show the nature of the whole.

Spectrochemical

Latin/Greek

specere- to look at, appearance

-khemeia a substance with a distinct molecular composition

Pertains to a series listing ligands based on their energy strengths; these differences cause different colors to be emitted.

Spectrophotometry

Greek

specere- to look at, appearance

-photos- light, radiant energy

-metria (metron) the process of measuring

The process of using an instrument to measure the intensity of various wavelengths of radiant energy.

Spectroscopy

Greek

specere- to look at, appearance

-scopium to look at, examine

Methods of studying substances exposed to some sort of exciting energy.

Spectrum

Latin

specere to look at, appearance

The distribution of energy emitted by a radiant source, as by an incandescent body, arranged in order of wavelength.

Speed

Old English

sped swiftness

The scalar quantity used to measure displacement per unit time.

Speleothem

Greek

spelaion cave

General name for any cave formation.

Sphenoid

Greek

sphen- wedge, wedge shaped

-oid (oeides) resembling; having the appearance of

The sphenoid bone or relating to the sphenoid bone; wedge shaped.

Sphincter

Greek

sphingein to bind tight

A ringlike muscle whose action resembles that of the drawstring of a bag. It normally serves to constrict an opening (mouth, anus, or arteriole) or, when relaxed, to enable access to the passage.

Spiracle

Latin

spir- breath of life, breath, breathing; mind, spirit, courage

-cle small

The external openings of the insect breathing (tracheal) system, found along the abdomen.

Spirochete

Greek/Latin

speira- coil

-chaeta bristle hair

Any of the various slender, spiral-shaped, motile bacteria.

Spirogyra

Greek

speira- coil

-guros ring

Any of various filamentous freshwater green algae of the genus *Spirogyra,* having chloroplasts in spirally twisted bands.

Spongocoel

Greek

spongos- sponge

-koilos hollow

Central cavity in sponges that opens to the exterior by an osculum.

Sporangium

Greek

sporos- seed

-angeion- vessel

-ium quality or relationship

Spore-containing structure; a sac or case in which spores are produced.

Sporophyte

Greek

spora- seed

-phuto plant

A stage in a plant's life cycle during which spores are produced.

Sporozoan

Greek

spora- seed

-zoan animal, animal-like

Member of the class Sporozoa, consisting of non-motile, single-celled parasitic organisms.

Stability
Latin
stabilis- to stand
-ity state of, quality of
Resistance to chemical change or to physical disintegration.

Stalactite
Greek
stalaktos- dropping or trickling
-ite minerals and fossils
An icicle-shaped, secondary mineral deposit that hangs from the roof of a cave.

Stamen
Latin
stamen thread
Reproductive, pollen-producing organ of a vascular plant, composed of a filament and an anther.

Staphylococcus
Greek
staphylo- cluster
-coccus of spherical or spheroidal shape
Spherical parasitic bacterium, usually occurring in grapelike clusters.

Static
Greek
statos- standing, stay, make firm, fixed
-ic (ikos) relating to or having some characteristic of
Of or relating to bodies at rest or forces that balance each other.

Stationary
Greek
statos- standing, stay, make firm, fixed
-ary of or relating to or connected with
Incapable of being moved, fixed; nonmotile organisms.

Statocyst
Greek
statos- standing, stay, make firm, fixed
-cyst (kustis) sac or bladder that contains fluid
A very small, fluid-filled organ found in many invertebrates that orients the body in relation to gravity.

Stearoptene
Greek
steat- fat, tallow
-ptenos volatile, winged
The more solid component of a volatile oil; it separates out as a whitish, crystalline solid as it cools to room temperature.

Steatohepatitis
Greek
steat- fat, tallow
-hepat- liver
-itis inflammation, burning sensation
Disease condition that is characterized by fatty deposits in the liver, that may or may not be caused by excessive alcohol use, and that has few symptoms that can be readily diagnosed.

Stegnosis
Greek
stegn- constriction, obstruction
-osis action, process, state, condition
A condition causing the stoppage of secretions; constriction, constipation.

Stegosaur
Greek
stegos- roof
-sauros lizard
Herbivorous dinosaur existing in the Jurassic to the Cretaceous periods and characterized by a double row of bony plates along the dorsal side, long rear legs, and a small head and neck.

Stele
Greek
stele pillar
The central core of tissue in the stem or root of a vascular plant.

Stenobenthic
Greek
stenos- narrow
-benth- deep; the fauna and flora of the bottom of the sea
-ic (ikos) relating to or having some characteristic of
Living within a narrow range at or near the bottom of the sea.

Stenocoriasis
Greek
stenos- narrow
-core- (corium) skin
-iasis a process or a pathological condition
The abnormal contraction of the pupil of the eye; a symptom of a pathological condition.

Stenocrotaphia
Greek
stenos- narrow
-crotaphion- pulse, beat
-ia names of diseases, place names, or Latinizing plurals
Narrowness of the temporal region.

Stenohaline

Greek

stenos- narrow

-halo- salt

-ine a chemical substance

Refers to organisms that are capable of tolerating only slight variations in salinity.

Stenothermal

Greek

stenos- narrow

-thermos- combining form of "hot" (heat)

-al of the kind of, pertaining to, having the form or character of

Describes an organism tolerant of only a narrow range of temperatures.

Stenothorax

Greek

stenos- narrow

-thoraces chest

Abnormal narrowness of the chest.

Stephanion

Greek

stephanos- crown

-ion state, process, or quality of

The point on the side of the cranium at which the coronal suture meets the superior temporal line.

Steradian

Greek

ster- solid

-radi- ray, spoke of a wheel

-an one that is of, relating to, or belonging to

Measurement of solid angles, equivalent to the angle subtended at the center of a sphere by an area on its surface equal to the square of its radius. A full sphere subtends 4π steradians.

Stereocilium

Greek

stereos- three-dimensional, solid, firm, hard

-cili- a small hair

-um (**singular**) structure

-a (**plural**) structure

A nonmotile protoplasmic filament on the free surface of a cell; found on hair cells of the inner ear and on pseudostratified epithelial cells of the male epididymis.

Stereopsis

Greek

stereos- three-dimensional, solid, firm, hard

-opisi vision

Stereoscopic vision allowing for depth perception and visual acuity.

Stereoscopic

Greek

stereos- three-dimensional, solid, firm, hard

-skopein- to view, examine

-ic (ikos) relating to or having some characteristic of

Pertaining to two images of the same scene, differing slightly in point of view, that are each seen by one eye, giving the effect of solidity.

Sternum

Greek

sternon- chest, breast, sternum, the breast bone

-um (**singular**) structure

-a (**plural**) structure

A long, flat bone articulating with the cartilages of the first seven ribs and with the clavicle, forming the middle part of the anterior wall of the thorax, and consisting of the corpus, manubrium, and xiphoid process.

Stethoscope

French/Greek

stethos- chest

-skopein to view, examine

Any of a group of instruments designed to amplify the sounds of the chest, such as heartbeat or respiration.

Stigma

Greek

stizein tattoo mark; to prick

A small pore, mark, or spot, such as the respiratory spiracle of an insect.

Stipule

New Latin

stipula trunk

Either of a pair of appendages borne at the base of the leafstalk in many plants.

Stoichiometry

Greek /English

stoicheious- element

-metria (metron) the process of measuring; to measure

A branch of science that deals with the application of the laws of definite proportions and of the conservation of matter and energy to chemical activity.

Stolon

Latin

stolo shoot

A shoot that bends to the ground or that grows horizontally above the ground, and that produces roots and shoots at the nodes.

Stomach
Greek
stomakhos gullet
The enlarged portion of the alimentary canal lying between the esophagus and the small intestine.

Stomata
Greek
stoma mouth
One of the minute pores in the epidermis of a leaf or stem through which gases and water vapor pass.

Stratigraphy
Latin
stratum- horizontal layer; stretched, spread out; layer, cloud layer
-graphia (graphein) to write, record, draw, describe
The study of the arrangement, distribution, and deposition of rocks in layers.

Stratosphere
Latin
stratum- horizontal layer; stretched, spread out; layer, cloud layer
-sphaire to surround
The second lowest layer of earth's atmosphere; the ozone layer is located in the upper stratosphere.

Stratovolcano
Latin
stratum- horizontal layer; stretched, spread out; layer, cloud layer
-vol'nus fire, flames (named after the Roman god of fire)
A volcano built up from alternating layers of rock and lava.

Stratus
Latin
stratum- horizontal layer; stretched, spread out; layer, cloud layer
Featureless sheets of clouds; horizontal, spread-out layers of grayish-colored clouds.

Strepsirhini
Greek
streptos- twisted chain, turn
-rhino nose, nasal
Suborder containing seven families of arboreal primates, formerly called prosimians, concentrated on Madagascar and having comma-shaped nostrils, a long nonprehensile tail, and a second toe provided with a claw.

Streptococcus
Greek
streptos- twisted chain, turn
-kokkos of spherical or spheroidal shape, grain, seed
Spherical bacteria that occur in pairs or chains.

Striation
Latin
stria- thin narrow groove or channels, bands
-ion state, process, or quality of
In biology, a group of protein bands found in skeletal muscle that are involved in muscular contractions. In earth science, one of a number of parallel lines or scratches on the surface of a rock that were inscribed by rock fragments imbedded in the base of a glacier as it moved across the rock.

Stromatolite
Greek
stroma- living on a bed; spread out
-lite combining form used in naming of minerals
Large mats and mounds composed of billions of photosynthesizing cyanobacteria that dominated the Proterozoic's shallow oceans.

Structure
Latin
structura part
A part of the body, such as the heart, a bone, a gland, a cell, or a limb.

Subcutaneous
Latin
sub- under or below
-cutis- skin
-ous full of, having the quality of, relating to
Refers to tissue or other object located just below the dermis or skin.

Subduction
Latin
sub- under or below
-ducere- to lead
-ion state, process, or quality of
Pertains to a long narrow zone associated with oceanic trenches, where one plate descends beneath another.

Sublimate
Latin
sublimus- up to, elevate, uplifted
-ate characterized by having
To purify or refine by subliming; to change matter from the solid state to the gaseous state or from the gaseous state to the solid state without an intervening liquid state.

Sublimation
Latin
sublimus- up to, elevate, uplifted
-ion state, process, or quality of
The process of changing a solid substance directly into a vapor without it first passing through the liquid state.

Sublime

Latin

sublimus up to, elevate, uplifted

To go directly from a solid to a gas without going through the liquid phase.

Subscript

Greek

sub- under or below

-scribere writing

A symbol written below another symbol or letter.

Substance

Latin

sub- under or below

-stantia- essence, material

-ance state, quality

A material produced by or used in a chemical process.

Subterranean

Latin

sub- under or below

-terra- earth

-an one that is of, relating to, or belonging to

Refers to that which is found beneath the earth's surface.

Succession

Latin

succedere- to follow after

-ion state, process, or quality of

The act of following in order; following consecutively.

Sugar

Middle English

sugre sugar

Any of various water-soluble compounds that vary widely in sweetness and include the oligosaccharides.

Supercell

Latin

super- superior in size, quality, number, or degree; exceeding the norm

-cella small room, compact, chamber

Self-sustaining, extremely powerful storm characterized by intense rotating updrafts.

Superconductivity

Latin

super- superior in size, quality, number, or degree; exceeding the norm

-conducere- to bring together

-ity state of, quality of

The flow of electric current without resistance in certain metals, alloys, and ceramics at temperatures near absolute zero, and in some cases at temperatures hundreds of degrees above absolute zero.

Supercooling

Latin

super- superior in size, quality, number, or degree; exceeding the norm

-cole- becoming less warm

-inde the act of

Cooling a liquid to a temperature below that at which crystallization would normally occur but without the separation of a solid.

Supernova

Latin

super- superior in size, quality, number, or degree; exceeding the norm

-nova new

A rare celestial phenomenon involving the explosion of most of the material in a star, resulting in an extremely bright, short-lived object that emits vast amounts of energy.

Surfactant

Old French

sur- above

-face- outward appearance

-agere to do

A surface-active substance designed to make a surface "wetter"; the fluid layer of the alveolar sacs of the lungs that makes the exchange of gases possible.

Susceptible

Latin

sus- (sub) below, under, beneath

-capere- catch, seize, take hold of, contain

-able/-ible capable, be inclined to, tending to, given to/capable of

Likely to be affected; permitting an action to be performed.

Suspension

Latin

suspendere- to cause to hang

-ion state, process, or quality of

A system consisting of a solid dispersed in a solid, liquid, or gas, usually in particles of larger than colloidal size.

Sustainable

Latin

sus- (sub) below, under, beneath

-tenere- to hold, grasp, have

-able capable, be inclined to, tending to, given to

Of, relating to, or being a method of harvesting or using a resource so that the resource is not depleted or permanently damaged.

Symbiosis

Greek

sym- with, together
-bios- life, living organisms, or tissue
-sis action, process, state, condition
The living together of two different species in an intimate relationship. The symbiont always benefits; the host may benefit, may be unaffected, or may be harmed (mutualism, commensalism, and parasitism, respectively).

Symmetrical

Greek

sym- with, together
-meter- (metron) instrument or means of measuring; to measure
-al of the kind of, pertaining to, having the form or character of
Regular as to the number of its parts; corresponding units of similar structure that exist on either side of a central axis.

Synapse

Greek

syn- together, united
-haptein- to fasten
-sis action, process, state, condition
Junction between two nerve cells, allowing the transfer of nerve impulses from the axon terminal of one neuron to another neuron or cell.

Synchronous

Greek

syn- together, united
-khronos- time
-ous full of, having the quality of, relating to
Occurring or existing at the same time; moving or operating at the same time.

Syncline

Greek

syn- together, united
-klinein to lean
A fold in rocks in which the rock layers dip inward from both sides toward the axis.

Syncytial

Latin

syn- together, united
-kutos- (cyto) sac or bladder that contains fluid
-al of the kind of, pertaining to, having the form or character of
Pertaining to a cytoplasmic mass that is multinucleated and lacks intercellular boundaries.

Syndiotactic

Greek

syndio- two together
-taktos ordered
Refers to the type of orientation of the methyl groups on a polypropylene chain in plastics—in this case alternating orientation.

Syndrome

Greek

syn- together, united
-dramein (dromos) to run
A group of signs and symptoms that occur together and characterize a particular abnormality.

Synecology

Greek

syn- together, united
-oikos- house
-logy (logos) used in the names of sciences or bodies of knowledge
Ecology of communities as opposed to individual species.

Synovial joint

Greek

syn- together, united
-ovo- egg
-ial (variation of *-ia*) relating to or characterized by
Freely moving joint in which two bones are separated by a cavity.

Synthesis

Greek

syn- together, united
-tithen- to put
-sis action, process, state, condition
The combining of separate elements or substances to form a coherent whole.

Systematics

Greek

syn- together, united
-histanai- set up
-ic (ikos) relating to or having some characteristic of
The systematic classification of organisms and the evolutionary relationships among them; taxonomy.

Systole

Greek

sustellein to contract
The rhythmic contractions of the ventricles of the heart that cause blood to be pumped from the heart into the aorta and the pulmonary arteries.

T

Tachycardia

New Latin

takhus- fast, swift

-kard- heart, pertaining to the heart

-ia names of diseases, place names, or Latinizing plurals

Faster than normal heart rate, usually calculated over 100 beats per minute in the resting state for adults.

Tachyon

English

takhus- fast, swift

-on a particle

A hypothetical subatomic particle that travels faster than the speed of light.

Tachypnea

Greek

takhus- fast, swift

-pnein breath

Breathing very rapidly.

Tarsal

Greek

tarsus- ankle

-al of the kind of, pertaining to, having the form or character of

A bone of the ankle; of or relating to the ankle.

Taxon (taxa)

Greek

taxis order, arrangement

Any taxonomic group or entity: kingdom, phylum, class, order, family, genus, or species.

Taxonomy

Greek

taxis- order, arrangement

-nom (nemein) to dictate the laws of, knowledge, usage, order

The classification of organisms in an ordered system that indicates natural relationships.

Technology

Greek

tekhne- skill, craft

-logy (logos) used in the names of sciences or bodies of knowledge

The application of science to situations usually, but not exclusively, associated with commerce and industry.

Tectonic

Greek

tekton- builder

-ic (ikos) relating to or having some characteristic of

In geology, relating to, causing, or resulting from structural deformation of the earth's crust. Study of the earth's structural features.

Telencephalon

Greek

tele- far off, distant

-enkephalos in the head

The anterior portion of the prosencephalon, constituting the cerebral hemispheres and composing with the diencephalon the prosencephalon.

Thomas Edison, the Great American Inventor

Few inventors in history were as prolific as Thomas Edison. When he was born, in 1847, the world was illuminated by candle and fire. When he died, in 1931, the world glowed in incandescent light. Though not his invention, he perfected the idea and came upon the necessary elements that would give light without burning out too soon.

Edison conducted most of his research at Menlo Park in New Jersey. There he would devote his life to producing some of the most widely used technology in history. Edison did not work alone. He had brilliant assistants with a single overriding objective: invent and produce. William Hammer, one of Edison's assistants, was the person in charge of perfecting the light bulb, and he did a remarkable job. In the year after the development of Edison's bulb, the Edison Lamp Works produced over 50,000 lamps.

Edison held 1,093 patents. With a steady flow of inventions, from his first patent ("Electrographic Vote-Recorder" in June 1869) to his last ("A Holder for Articles to Be Electroplated," submitted in May 1933), Edison and his assistants invented and patented such gadgets as the printing telegraph, the electric switch, electromagnetic telegraphic instruments, the typing wheel for telegraphs, the galvanic battery, the speaking machine, the phonograph, the vacuum pump, the electric generator, the typewriter, the electric meter, the electric indicator, the electric railway, the electrical transmission of power, phonogram blanks, the motion picture camera, railway signaling, the voltaic battery, the electric locomotive, the magnetic separator, the gas purifier, the cement kiln, an electronic system for automobiles, a process for constructing concrete buildings, improvements to the telephone, and on and on.

Thomas Edison died in 1931. He, along with a few other men in his lab, changed American society forever. Through his inventions and his strong business sense, he managed to get his inventions manufactured at a cost that was affordable to many. In a tribute to his passing, the lights were dimmed for one minute on October 21, 1931, a few days after his death.

Telescope
Greek
tele- far off, distant
-skopos watcher
An optical instrument used for viewing distant objects by means of the refraction of light rays through a lens.

Telophase
Greek
telos- end
-phasis appearance
The final of the four stages of nuclear division in mitosis and each of the two divisions in meiosis.

Telson
Greek
telson limit
The rearmost segment of the body of certain arthropods; an extension of this segment, such as the middle lobe of the tail fan of a lobster or the stinger of a scorpion.

Tendon
Greek
tenon- tendon, sinew, to stretch
A band of tough, inelastic fibrous tissue that connects a muscle with its bony attachment.

Tenodesis
Greek
tenon- tendon, sinew, to stretch
-desis binding, fixation
The surgical fixation of a tendon to a bone.

Tenoplasty
Latin/Greek
tenon- tendon, sinew
-plastos (plassein) something molded (to mold)
Reparative or plastic surgery of the tendons.

Tension
Latin
tension- an extension or length
-ion state, process, or quality of
A force supplied by a rope or chain whose direction is away from the load.

Tentacles
Latin
tentare to feel, try
A flexible extension, such as one of those surrounding the mouth or oral cavity of the squid, used for feeling, grasping, or locomotion.

Tephra
Greek
tephra ash
The solid substance ejected from a volcanic eruption.

Teratological
Greek
terat- marvel, omen, monster
-logo- talk, speak
-al of the kind of, pertaining to, having the form or character of
Monstrous, relating to monstrosity; the biological study of birth defects.

Terrain
Latin
terrenus of the earth
A series of related rock formations.

Tertiary
Latin
tertius- third
-ary of, relating to, or connected with
First period of the Cenozoic era, extending from the beginning of the Paleocene epoch over 58 million years ago to the end of the Pliocene epoch 2 million years ago.

Tetrad
Greek
tetras four
A group or set of four homologous chromosomes.

Tetrahedron
Greek
tetra- four faced
-hedron head
A polyhedron with four faces; a Platonic solid P5.

Thallophytes
Greek
thallos- young green shoot
-phyte a plant
A major group of organisms formerly belonging to the plant kingdom. They lack true roots, stems, and leaves. Representative samples include algae, fungi, and mosses.

Thallus
Greek
thallos- young green shoot
-us thing
A plant that possesses an undifferentiated stem and lacks true vascular tissue.

Thermoacidophile
Greek
thermos- combining form of "hot" (heat)
-acido- of or related to an acid
-phile one who loves or has a strong affinity or preference for
An organism that thrives in a strongly acidic environment at high temperatures.

Thermocline
Greek
thermos- combining form of "hot" (heat)
-klinein to lean, sloping
The transitional layer between warm surface waters and the cold bottom water of oceans or lakes.

Thermodynamic
Greek
thermos- combing form of "hot" (heat)
-dynamique- powerful
-ic (ikos) relating to or having some characteristic of
Characteristic of or resulting from the conversion of heat into other forms of energy.

Thermograph
Greek
thermos- combining form of "hot" (heat)
-graphia (graphein) to write, record, draw, describe
A thermometer that records temperatures independently of humans by graphing the data on paper or recording the data electronically.

Thermometer
Greek
thermos- combining form of "hot" (heat)
-meter (metron) instrument or means of measuring; to measure
A device usually consisting of a graduated glass tube filled with either alcohol or mercury that is used to measure temperature.

Thermophile
Greek
thermos- combining form of "hot" (heat)
-phile one who loves or has a strong affinity or preference for
Any group of organisms that have adapted to and thrive in environments of extreme heat, usually over 45 degrees Celsius.

Thermosphere
Greek
thermos- combining form of "hot" (heat)
-sphaira a globe shape, ball, sphere
The outermost layer of the earth's atmosphere.

Thermostat
Greek
thermos- combining form of "hot" (heat)
-statos standing, stay, make firm, fixed, balanced
An automatic device for regulating temperature.

Thigmotropism
Greek
thigma- to touch
-trope- bend, curve, turn, a turning; response to stimulus
-ism state or condition, quality

The turning or bending response of an organism upon direct contact with a solid surface or object.

Thoracic
Greek
thorakikos- thorax, chest
-ic (ikos) relating to or having some characteristic of
Of, pertaining to, or situated in or near the chest.

Thoracocentesis
Latin
thorakikos- thorax, chest
-cente- puncture
-sis action, process, state, condition
Aspiration of the pleural cavity. A surgical procedure where the chest wall is punctured to allow for the drainage of fluids from the chest.

Thorax
Greek
thorakikos thorax, chest
The cage of bone and cartilage where the primary organs of the respiratory system reside. Formed ventrally by the sternum and costal cartilages and dorsally by the twelve thoracic vertebrae connected to the dorsal parts of the twelve ribs.

Thrombocyte
Greek
thrombo- clot, blood clot
-cyte (kutos) sac or bladder that contains fluid
A cell, specifically platelets responsible for initiating the clotting of blood.

Thrombocytopenia
Greek
thrombo- clot, blood clot
-kutos- (cyto) sac or bladder that contains fluid
-penia reduction, poverty, lack, deficiency
A reduced number of platelets in the blood.

Thrombosis
New Latin
thrombo- clot, blood clot
-sis action, process, state, condition
Formation of a clot in a blood vessel.

Thrust
Old Norse
thrysta to tire
The force provided to drive an object through a medium, such as an airplane through air.

Thylakoids
Greek
thylakos- sack
-oid (oeides) resembling, having the appearance of
Fattened sac within a granum whose membrane contains chlorophyll and where the light-dependent reactions of photosynthesis occur.

Thymine
Greek
thym(ic) acid- acid from the thymus
-ine of or relating to
An essential nitrogenous base found in DNA.

Thymus
Greek
thumos wartlike outgrowth
A tiny lymphatic gland located behind the sternum. It is active in young people and is mostly involved with T cell differentiation. It diminishes in size and becomes vestigial in adults.

Thyroid
Latin
thureos- oblong shield, door
-oid (oeides) resembling; having the appearance of
An endocrine gland located laterally to the trachea in mammals; it produces various hormones, including triiodothyronine and calcitonin.

Thysanoptera
Greek
thysanos- fringe
-pteron feather, wing
An insect order classified as being minute to small, with long, narrow bodies and broadly fringed wings (also know as thrips).

Thysanura
Greek
thysanos- fringe
-ura tail
Silverfish; wingless, quick-moving, flattened insects that lack metamorphosis and are considered by humans to be a pest species.

Tide
Old English
tima division of time
The periodic variation in the surface level of the oceans caused by the gravitational attraction of the moon and the sun.

Time
Anglo Saxon
tima time, hour, or season
The period between two events.

Tinnitus
Latin
tinnire to ring
A ringing sound in the ears, the cause of which is unknown.

Titrate
French
titre- concentration of a substance
-ate characterized by having

To determine the concentration of a substance by titration.

Titration
Latin
titre- concentration of a substance
-ion state, process, or quality of
The process of determining the concentration of a substance in solution by adding to it a standard reagent of known concentration in carefully measured amounts until a reaction of definite proportion is completed.

Tongue
Latin
tunge tongue
A muscular organ that is usually attached to the floor of the mouth.

Tonsil
Latin
toles tonsil
Mass of lymphoid tissue in the back of the mouth and the throat and on the rear of the tongue.

Topography
Greek
topos- place
-graphia (graphein) to write, record, draw, describe
The configuration of a surface, including its relief and the position of its natural and man-made features.

Torque
Latin
torquere to twist
The moment of a force or the measurement of a force's tendency to produce torsion or rotation around an axis.

Toxic
Greek
toxikos- poison
-ic (ikos) relating to or having some characteristic of
Having to do with poison or something harmful to the body.

Toxicity
Greek
toxikos- poison
-ity state of, quality of
Of, relating to, or caused by a poison or toxin.

Toxicomania
Greek/English
toxikos- poison
-mania obsessive preoccupation with something; madness, frenzy; obsession, or abnormal desire for
An intense craving for poisons; an urge to poison oneself.

Trachea
Greek/Latin
trakheia rough
Main trunk of the system of tubes by which air passes to and from the lungs.

Trait
Latin
tractus drag, drawing out, line
A distinguishing quality; an inherited characteristic.

Trajectory
Latin
traicere- to cause to cross.
-ory of or pertaining to
The path followed by a projectile.

Transcription
Latin
trans- across or through
-scribere to write down
A process in which DNA serves as a template for RNA formation.

Transduction
Latin
transducere- transfer
-ion state, process, or quality of
The transfer of genetic material from one microorganism to another by a viral agent.

Transfer
Latin
trans- across or through
-ferre to carry
To convey or cause to pass from one place, person, or thing to another.

Transformation
Latin
trans- across or through
-forma- shape
-ion state, process, or quality of
The alteration of a bacterial cell caused by the transfer of DNA from another bacterial cell.

Transfusion
Latin
trans- across or through
-fundere- to pour
-ion state, process, or quality of
The act of instilling, moving, or transferring a substance from one vessel to another.

Transgenesis
Latin
trans- across or through
-gen- to give birth, kind, produce
-sis action, process, state, condition

Integration into a living organism of a foreign gene that confers upon the organism a new property that it will transmit to its descendants.

Transgenic
Latin
trans- across or through
-gen- to give birth, kind, produce
-ic (ikos) relating to or having some characteristic of
Refers to an organism that contains genes from another species, where the genes contain foreign DNA.

Translation
Latin
trans- across or through
-latus- brought
-ion state, process, or quality of
The process by which mRNA directs the amino acid sequence of a growing polypeptide during protein synthesis.

Translocation
Latin
trans- across or through
-locus- place
-ion state, process, or quality of
The rearrangement of genetic material within the same chromosome, or the transfer of a segment of one chromosome to another, nonhomologous one.

Translucent
Latin
trans- across or through
-lucere- to shine
-ent causing an action, being in a specific state, within
Transmitting light but causing sufficient diffusion to prevent the perception of distinct images.

Translunar
Latin
trans- across or through
-luna- moon
-ar relating to or resembling
Extending beyond the moon or its orbit around the earth.

Transmission
Latin
trans- across or through
-miss- to let go or to send
-ion state, process, or quality of
The process of causing to pass through, be conveyed, or be sent out.

Transpiration
Latin
trans- across or through
-spir- to breathe
-ion state, process, or quality of
The evaporative loss of water from a plant.

Transplant
Latin
trans- across or through
-plantare to plant
To uproot a plant from one area to another, or to remove an organ or tissue from an animal and place it in another.

Transport
Latin
trans- across or through
-portare carry
The movement or transference of biochemical substances from one site to another.

Transverse
Latin
trans- across or through
-vertere to turn
Situated or lying across; crosswise.

Trematode
Greek
tremat- perforation
-hodos wave
A class of parasitic flatworms that attach themselves to hosts by hooks or suckers.

Triassic
Latin
trias- three
-ic (ikos) relating to or having some characteristic of
Of or belonging to the geologic time, system of rocks, or sedimentary deposits of the first period of the Mesozoic era, characterized by the diversification of land life, the rise of dinosaurs, and the appearance of the earliest mammals.

Triboluminescence
Greek/Latin
tribein- to rub
-lumen- light
-ence the condition of
The production of light taking the appearance of tiny sparks that are observed in the dark in some minerals when a hard point is dragged across the surface of the mineral.

Triceps
Latin
tri- three
-caput head
A muscle with three points of origin.

Triceratops

Greek
tri- three
-keras- horn
-ops eye, face
A herbivorous dinosaur of the genus *Triceratops*, of the Cretaceous period, having a bony plate covering the neck, a large horn above either eye, and a small horn on the nose.

Trichinella

Greek
trichinos- made of hair
-ella little
One of the group of parasitic nematodes that are slender and hairlike; roundworms that cause trichinosis.

Trichocyst

Greek
trichinos- made of hair
-cyst (kustis) sac or bladder that contains fluid
A threadlike stinging or grasping structure possessed by some ciliates and other protists that is used for capturing prey.

Trichoptera

Greek
trichino- made of hair
-pteron feather, wing
The four-winged insect order whose species are found near lakes and streams; caddisflies.

Trichroism

Greek
tri- three
-khros- color
-ism state or condition, quality
The property possessed by certain minerals in which three different colors are displayed when the mineral is viewed from three different directions under white lights.

Triclinic

Greek
tri- three
-klinein to lean, sloping
Having three unequal axes intersecting at oblique angles.

Tricuspid

Greek
tri- three
-cuspis- sharp point, cusp
-id state, condition; having, being, pertaining to, tending to, inclined to
Structure having three cusps; the molars (teeth) and the tricuspid valve of the human heart.

Trigeminal

Greek
tri- three
-gemin- twin, double
-al of the kind of, pertaining to, having the form or character of
The main sensory nerve of the face and motor nerve for the muscles of mastication.

Trisomy

Greek
tri- three
-soma- (somatiko) body
-y place for an activity, condition, state
Abnormal condition of having three copies of a chromosome rather than the normal two in a somatic cell.

Trophozoite

Greek
trophos- (trophein) to nourish, food, nutrition; development
-zoion animal, living being
The adult, active feeding stage of unicellular organisms in the class Sporozoa.

Tropism

Greek
trope- bend, curve, turn, a turning; response to stimulus
-ism state or condition, quality
The turning or bending movement of an organism toward or away from an external stimulus.

Tropopause

Greek
trope- bend, curve, turn, a turning; response to stimulus
-pausis stop
Atmospheric region between the troposphere and the stratosphere.

Troposphere

Greek
trope- bend, curve, turn, a turning; response to stimulus
-sphaira a globe shape, ball, sphere
The lowest region of the atmosphere between the earth's surface and the tropopause, characterized by decreasing temperature with increasing altitude.

Trough

Middle English
trog wooden vessel
The minimum point in a wave or alternating signal.

Tsunami

Japanese

tsu- port
-nami wave
A large ocean wave caused by an underwater earthquake or volcanic eruption.

Tubule
Latin
tubus- pipe
-ule little, small
A very small tube or tubelike structure.

Tufa
Latin
tufos tuff
Calcareous lime deposits usually formed as precipitates from springs with high concentrations of calcium; unusual formations of lime deposits.

Tumor
Latin
tumere to swell
An abnormal growth of tissue characterized by a proliferation of cells serving no useful purpose.

Tympanic
Greek
tumpanon- drum
-ic (ikos) relating to or having some characteristic of
Relating to the membrane, a diaphragm-like structure that is external on some insects and internal in mammals.

Tyrannosaur
Greek
turannos- tyrant
-sauros lizard
A large dinosaur with small forelimbs, a large head, and a strong tail that existed during the Upper Cretaceous period in North America.

Ulcer
Latin
ulcus open sore
Lesion of the skin or mucous membrane in which bleeding usually occurs and necrosis of the surrounding tissue often occurs.

Ultraviolet
Latin
ultra- beyond, to an extreme degree
-violet shortest ray on the visible spectrum
Lying just beyond the violet end of the visible spectrum.

Umbra
Latin
umbra shadow
The completely dark portion of the shadow cast by the earth, moon, or other body during an eclipse.

Undifferentiated
Latin
un- not
-differens different
Refers to cells during embryonic growth that have not yet developed into organs and tissues with specialized functions.

Ungulate
Latin
unguis- hoofed, clawed, nail
-ate characterized by having
Of or belonging to the former order Ungulata; hooved mammals such as horses, cattle, deer, and swine.

Unicellular
Latin
uni- same, one
-cellul- cell, small room
-ar relating to or resembling
Plant and animal-like organisms that have or consist of one cell; to be one-celled.

Uniform
Latin
uni- same, one
-forma shape
Being always the same, as in character or degree; unvarying.

Uniparous
Latin
uni- same, one
-para- to bring forth, to bear
-ous full of, having the quality of, relating to
Refers to animals that produce one offspring at a time or to plants that form a single axis at each branching.

Unit
Latin
unus one
A determinate quantity adopted as a standard of measurement.

Unsaturated
Latin
un- not
-satur- full
-ate characterized by having
Containing less of a solute required for equilibrium.

Uracil

Latin
urina- (ur)ea urine
-acetum- (ac)ectic acetic acid, vinegar
-il substance relating to
An essential chemical of RNA.

Urease

Latin
urea- urine
-ase enzyme
An enzyme that promotes the hydrolysis of urea.

Ureter

Greek
ouron- water, rain, wet; urine
-ter denoting the instrument
A thick-walled tube that conveys urine from the kidney to the urinary bladder.

Urethra

Greek
ourethra urinate
A canal extending from the bladder to the exterior of the body; it carries urine in both sexes and semen in males of the species.

Urinary

Greek
ouron- water, rain, wet; urine
-ary of, relating to, or connected with
Of or relating to the organs involved in the formation and excretion of urine.

Uropod

Greek
uro- tail
-pod foot
One of the abdominal appendages of a crustacean, which are used chiefly in locomotion.

Uterine

Latin
uterus- womb
-ine of or relating to
Of, pertaining to, or in the region of the uterus.

Uterus

Latin
uterus womb
A hollow muscular organ of the female mammal for the gestation of fetuses, located in the pelvic region.

Utilization

Latin
utilize- to use
-ion state, process, or quality of
The act or process of putting something to use for a productive purpose.

Uvula

Latin
uva- grape (swollen)
-ula little, small
A small, pendant/grape-shaped, fleshy mass of tissue suspended from the center of the posterior border of the soft palate.

V

Vaccine
Latin
vacc- cow
-ine a chemical substance
A substance prepared from pathogens that is injected into the body in order to build antibodies and create immunity from diseases caused by those pathogens.

Vacuole
Latin/French
vacuus- empty
-ole little
A membrane-enclosed cavity that contains water, food, or wastes from cellular activity.

Vagina
Latin
vagina sheath
A tube or canal that extends from the uterus to the exterior of the body.

Valence
Latin
valere to be strong
Any number given to an element or ion as an indicator of combining sites; also used to determine whether electrons will be gained or lost as a result of a chemical reaction.

Vapor
Latin
vapor diffuse matter in air
Suspended liquid, particulate matter, or smoke within a gas, such as steam or fog.

Vaporization
Latin
vapor- diffuse matter in air
-ize to make, to treat, to do something with
-ion state, process, or quality of
The process of converting a liquid into a gas.

Vaporize
Latin
vapor- diffuse matter in air
-ize to make, to treat, to do something with
To convert or be converted into vapor.

Variation
Latin
variare- different, diversity, change
-ion state, process, or quality of
Divergence in the characteristics of an organism from the species or population norm or average.

Varicose
Latin
varic- swollen vein
-ose full of, containing, having the qualities of, like
Describes the abnormal condition of swollen or twisted superficial veins.

Variegation
Latin
varius- various
-agere- to do, drive
-ion state, process, or quality of
Irregular variation in the color of plant organs, such as leaves or flowers.

Vas deferens

Latin

vas- vessel, duct

-de- reverse the action of, undo, from, apart, away

-ferre to carry

The duct or tubule by which semen is carried from the epididymis to the ejaculatory duct.

Vascular

Latin

vas- vessel, duct

-cul- small, tiny

-ar relating to or resembling

Characterized by containing vessels that carry or circulate fluids through plants and animals.

Vasodilation

Latin

vas- vessel, duct

-di- apart, away, from

-latus- wide

-ion state, process or quality of

The act or process of increasing the diameter of a small blood vessel.

Vasopressin

Latin

vas- vessel, duct

-premere- to press, curtail, prohibit

-in protein or derived from a protein

Antidiuretic hormone (ADH) secreted by the anterior lobe of the pituitary gland. This hormone simultaneously constricts small blood vessels, raises blood pressure, and reduces urinary output.

Vasospasm

Latin/Greek

vas- vessel, duct

-spasmos involuntary contraction, pull

Constriction of a blood vessel.

Vastus

Latin

vastus broad, large

Term suggesting "large" or "broad," in reference to muscle size.

Vector

Latin

vehere to carry

In physics, a quantity with both magnitude and direction. In biology, an organism that carries pathological organisms and delivers them from one host to another. In genetics, a plasmid or other agent that carries genetic material from one cell to another.

Vegetation

Latin

vegetat- to enliven

-ion state, process, or quality of

The act or process of vegetating; plants growing in a given area.

Vein

Latin

vena vessel, tube

Large blood vessel that conducts blood toward the heart.

Velocity

Latin

velox- quick

-ity state of, quality of

The vector quantity used to measure speed.

Vena cava

Latin

vena- vein

-cava empty, hollow

Very large veins, both superior and inferior, that empty blood into the right atrium of the heart.

Vent

Latin

ventus wind

The opening of a volcano in the earth's crust.

Ventifact

Latin

ventus- wind

-(arti)fact product or result

A stone that has been shaped by wind-driven sand.

Ventral

Latin

venter- belly

-al of the kind of, pertaining to, having the form or character of

Of or close to the abdomen, on the front of the human body or on the lower side of an animal or fish.

Ventricle

Latin

ventricul- belly

-us thing

One of the small chambers or cavities usually associated with the heart or brain.

Venule

Latin

vena- vessel, tube

-ule little, small

Smaller blood vessel that conducts blood toward a larger vein that ultimately returns blood to the heart.

Vermiculite

Latin
vermis- worm
-lithos- stone, rock
-ite minerals and fossils
Any of a group of micaceous hydrated silicate minerals related to the chlorites and used in heat-expanded form as insulation and as a planting medium.

Vermiform

Latin
verm- worm
-forma having the form of
A legless, wormlike larva without a well-developed head.

Vertebrate

Latin
vertebratus- jointed
-ate characterized by having
Having a backbone or spinal column; an animal in the phylum Chordata, subdivision Vertebrata.

Vertex

Latin
vertere to turn
The point at which the sides of an angle intersect; the highest peak of a mountain.

Vertical

Latin
vertic- highest point
-al of the kind of, pertaining to, having the form or character of
The axis perpendicular to the horizon (up and down); positioned at the highest point.

Vertigo

Latin
vertere to turn
The sensation of a whirling or spinning motion associated with oneself or with external objects; confused or disoriented.

Vesicle

Latin
vesic- little bladder
-ula little, small
Within the cytoplasm of cells, one of a variety of small, membrane-bound sacs that function in the transport, storage, or digestion of substances or in some other activity.

Vestigial

Latin
vestigium- no sign of any return
-ial relating to or characterized by
Refers to an indication, either by structural feature or some other minute piece of evidence, of the existence of a body part that no longer is present in the modern species (i.e., the forelimbs of ostriches).

Vibration

Latin
vibrare- to move back and forth
-ion state, process, or quality of
The act or process of rapidly moving back and forth.

Vibrissae

Latin
vibro- to quiver, to oscillate
-ae plural
Stiff hairs or feathers, usually projecting from the face (i.e., whiskers).

Villus

Latin
vill- tuft of hair or fleece
-us thing
Small, fingerlike projections extending into the interior of the small intestine and increasing the absorptive area of the intestinal wall.

Viper

Latin
vipera snake
Any of several venomous Old World snakes of the family Viperidae, having a single pair of long, hollow fangs and a thick, heavy body.

Viremia

Latin
virus- poison
-emia the condition of having (a specific thing) in the blood
Viruses found moving within the bloodstream; they may be pathogenic.

Virus

Latin
virus poison
Any of various simple submicroscopic parasites of plants, animals, and bacteria that often cause disease.

Visceral

Latin
viscidus- sticky
-al of the kind of, pertaining to, having the form or character of
Of the internal organs of the body, such as the heart, lungs, and intestines.

Viscosity

Latin
viscosus- sticky
-ity state or quality

Numerical measure of the degree to which a fluid resists flow under an applied force.

Vision

Latin

videre- to see

-ion state, process, or quality of

Eyesight; the ability to see.

Vitamin

Latin

vita- live

-ammonia- a colorless pungent gas, NH_3

-ine a chemical substance

Various water- or oil-soluble organic substances that are ingested in small amounts and are essential for growth and development.

Vitreous

Latin

vitrium- glass

-ous full of, having the quality of, relating to

Of or resembling glass; clear substance.

Viviparity

Latin

viva- life, alive

-para- to bring forth, to bear

-ity state of, quality of

Reproduction in animals whose embryos develop within the female parent and derive nourishment from her tissues (i.e., the placenta).

Volatile

Latin

volare- to fly

-ile changing, ability, suitable, tending to

Refers to that which readily evaporates at room temperature and pressure.

Volcanic

Latin

vol'nus- fire, flames (named after the Roman god of fire)

-ic (ikos) relating to or having some characteristic of

Pertains to extrusive rocks that cool above the surface.

Volcano

Latin

vol'nus fire, flames (named after the Roman god of fire)

A mountain formed of lava, ash, and larger fragments ejected during numerous eruptions.

Volume

Latin

volumen to roll

The amount of space occupied by a three-dimensional object or region of space, expressed in cubic units.

Volvox

Latin

volvere to roll

Hollow, spherical, multicellular green algae of the genus *Volox* that are found in freshwater.

Vulva

Latin

vulva womb, covering

The external genitalia of the female, including the labia, hymen, perineum, and clitoris.

Water
Old English
wæter water
Odorless, colorless, tasteless fluid vital to all plants and animals.

Wattle
Old English
watel hurdle
A fleshy, wrinkled, often brightly colored process hanging from the neck or throat, common in certain birds, such as chickens.

Wax
Old English
weax wax
Oils and greases composed of hydrocarbons and esters that are quite sensitive to heat and insoluble in water.

Weather
Old English
weder weather
The regional condition of the atmosphere with respect to temperature, humidity, precipitation, and wind.

Weight
Old English
wegan to weigh
The force on an object as a result of gravitation.

Work
Greek
ergon activity
The amount of energy required to exert a force over a given distance.

Henry Cavendish

Perhaps Henry Cavendish lost his chance at fame and glory because of his odd, quirky personality. Henry was painfully shy toward strangers and women. He was, however, respected and admired by his colleagues. According to accounts from his contemporaries, Henry would refrain from making eye contact with anyone but those closest to him.

Henry Cavendish was born in Nice, France, on October 10, 1731, and he died 78 years later, on February 24, 1810. During his sequestered life, Henry discovered some of the most important prin-

ciples of chemistry but historically has been given little credit for those discoveries. After his death, many of Cavendish's discoveries were later made by others. It wasn't until James Clerk Maxwell, a Scottish mathematician, went through Cavendish's writings in the latter part of the nineteenth century that the outside world realized what Henry had accomplished in his life. Ohm's law, Dalton's law of partial pressure, and Charles' law of gases, though not so named, were among the principles of chemistry included in Cavendish's narratives.

By experimentation, Cavendish was able to accurately calculate the density of the earth relative to water. The results of his experiments led to the calculation of the actual mass of the earth. He was accurate to within 1 percent of the earth's actual mass, which is estimated at 5.9725 billion trillion tons.

We associate Henry Cavendish with the discovery of the composition of water. Cavendish is given credit for the discovery of hydrogen, although, again, he didn't name it as such. That did not happen until Antoine Lavoisier researched Cavendish's experiments in 1777 and carried on with them.

Henry Cavendish's experiments with gases were meticulously conducted. He repeated his trials with gases over and over as he attempted to successfully differentiate them by their specific gravity.

Cavendish accurately established the composition of earth's atmosphere as being 79.167 percent "phlogisticated" (inflammable) air and 20.8333 percent "dephlogisticated" air. Today we know that most of the phlogisticated air is nitrogen and the dephlogisticated air is oxygen.

dephlogisticated air + inflammable air →water

[Now: $2 H_2 (g) + O_2 (g) \rightarrow H_2O (l)$]

Xanthic
Greek
xanthos- yellow
-ic (ikos) relating to or having some characteristic of
In botany, pertains to any plant or fruit that has a tendency to be yellowish in color.

Xanthophyll
Greek
xenos- stranger, different
-phyll leaf
Yellow pigment that is found in the leaves of green plants and is masked by the green pigment chlorophyll.

Xenobiotic
Greek
xeno- guest
-bios- life, living organisms, or tissue
-ic (ikos) relating to or having some characteristic of
Pertains to a drug or other foreign substance capable of harming another living thing.

Xenocrystal
Greek
xenos- stranger, different
-krustallos ice
A crystal foreign to the igneous rock in which it occurs.

Xenogenic
Greek
xenos- stranger, different
-gen- to give birth, kind, produce
-ic (ikos) relating to or having some characteristic of
Refers to a trait originating from a genetically different species and introduced into an organism.

Xenotransplantation
Greek/Latin
xenos- stranger, different
-trans- across or through
-plantare- to plant
-ion state, process, or quality of
The surgical removal of an organ or tissue from one species and the transplantation of it into a member of a different species.

Xerophyte
Greek
xeros- dry, arid
-phyte plant
A plant that lives in dry ecosystems, such as deserts.

Xiphoid
Greek
xiphos- sword
-oid (oeides) resembling; having the appearance of
Refers to the pointed, cartilaginous tip attached to the lower end of the breastbone or sternum; the smallest and lowest division of the sternum.

Xylem
Greek
xulon wood
The supporting and water-conducting tissue of vascular plants, consisting primarily of woody tissue.

Xylophage
Greek
xulon- wood
-phage eat, eating, consume, ingest
An organism that eats wood, typically an insect. Certain mollusks and fungi also bore into wood.

Y

Yeast
Old English
gist yeast
Single-celled fungi belonging to the families Ascomycetes and Basidiomycetes.

Yew
Old English
iw yew
A type of evergreen tree found mostly in temperate climates and thriving in acid soils.

Yield
Old English
gelda to pay

In biology, the amount of food gathered from a given crop. In chemistry, the amount of product obtained from a given chemical reaction.

Yolk
Old English
geolu yellow
The yellow substance of an egg, composed of water, protein, and lipids, that is surrounded by a clear, proteinatous layer of albumen.

Youze
East India
youze cheetah
The cheetah.

Z

Zeatin
Greek
zeia- wheat, barley, corn
-in protein or derived from a protein
A plant hormone found in the endosperm of maize fruits.

Zein
Greek
zeia wheat, barley, corn
A protein found in corn that is used in plastics, coatings, and adhesives

Zenith
Latin/Arabic
semita path over the head
The point on the celestial sphere that is directly above the observer.

Zeolite
Greek
zein- to boil
-lithos rock, stone
Aluminum silicate mineral whose molecules enclose cations of sodium, potassium, calcium, strontium, or barium; used chiefly as molecular filters and ion-exchange agents.

Zero
Arabic
sifr nothing, cipher
Empty, nothing; the absence of any integer.

Zinc
Old German
zinko spiked (because it became spiked or jagged in the oven)

A metal that is whitish in color and malleable at warm temperatures; one of a group of metals used in the making of alloys.

Zircon
Persian
zargun- (Persian form **āzargūn**) gold colored
āc- (as in **āçiyādiya**) fire worship month
-gūn color
Stable mineral found in granite and that provides evidence for the earth's crust being at least 4.2 billion years old; a brown to colorless mineral, $ZrSiO_4$, which is heated, cut, and polished to form a brilliant, blue white gem.

Zoanthropy
Greek
zoon- animal, animal-like
-anthropo- man; human being, mankind
-y place for an activity, condition, state
A mental disorder categorized as a monomania, where an individual believes he has transformed himself into another animal.

Zone
Greek
zone girdle, celestial zone
A distinctive region or area that is characterized by a common set of features and relatively distinct boundaries.

Zoobenthos
Greek
zoon- animal, animal-like
-benthos deep; the fauna and flora of the bottom of the sea
Those fauna living in or on the seabed or lake floor.

Zoodomatia
Greek
zoon- animal, animal-like
-domatia commune, home
Plant structures that act as shelters for animals.

Zooflagellates
zoon- animal, animal-like
-flagell- a whip
-ate characterized by having
A group of animal-like protists that are characterized by having flagella.

Zoology
Greek
zoon- animal, animal-like
-logy (logos) used in the names of sciences or bodies of knowledge
The branch of biology that deals with the study of the structure, physiology, development, and classification of animals.

Zoonosis
Greek
zoon- animal, animal-like
-noso- disease
-sis action, process, state, condition
Any infection of a human by a pathogen whose source is a reservoir of a nonhuman animal pathogen.

Zooparasite
Greek
zoon- animal, animal-like
-para- beside; near; alongside
-sitos- grain, food
-ite resident
An animal that feeds off a host organism.

Zoophagous
Greek
zoon- animal, animal-like
-phagos- (phagein) to eat, eating
-ous full of, having the quality of, relating to
A broad term applied to animals that feed off other animals.

Zoophyte
Greek
zoon- animal, animal-like
-phyte a plant
Any animal that resembles a plant more than an animal in morphology or mode of life.

Zooplankton
Greek
zoon- animal, animal-like
-planktos- passively drifting, wandering, roaming
-on a particle
Small animals that float or swim near the surface of water.

Zooplasty
Greek
zoon- animal, animal-like
-plastos- (plassein) something molded; to mold
-y place for an activity, condition, state
The surgical procedure whereby animal tissue is grafted and implanted in humans.

Zoosmosis
Greek
zoon- animal, animal-like
-osmos- for thrust, push
-sis action, process, state, condition
The osmotic process occurring in living systems, specifically in animals.

Zoosporangium
Greek
zoon- animal, animal-like
-spora- seed
-y place for an activity, condition, state
A vesicle in plants that holds zoospores.

Zoospore
Greek
zoon- animal, animal-like
-spora seed
Spores possessing flagella that are capable of locomotion.

Zootoxin
Greek
zoon- animal, animal-like
-toxicum poison
A poison produced by an animal.

Zooxanthella
Greek
zoon- animal, animal-like
-xanthos- yellow
-ella dimunitive
Microscopic yellow-green algae that live symbiotically within the cells of coral.

Zwitterion
German
zwitterion hybrid ion
A molecule that has positive and negative charges on opposite sides; a dipolar molecule.

Zygodactylous
Greek
zugon- to yoke, pair
-daktulos- toe, finger, digit
-ous full of, having the quality of, relating to
A term applied to yoke-toed birds such as woodpeckers, parrots, and cuckoos; the toes of these

birds are in sets of two, with one set lying anterior to the leg and the other posterior.

Zygoma
Greek
zugoun to join, bolt
The slender bony arch that joins the cheek to the temporal bone.

Zygomatic
Greek
zugoun- to join, bolt
-ic (ikos) relating to or having some characteristic of
Of or relating to the area of the zygoma.

Zygomorphic
Greek
zugon- to yoke, pair
-morph- shape, form, figure, or appearance
-ic (ikos) relating to or having some characteristic of
Refers to an organism having a paired or bilateral symmetry.

Zygospore
Greek
zugon- to yoke, pair
-spora seed, a sowing
A thick-walled spore of some algae and fungi formed by the union of two similar sexual cells; usually serves as a resting spore and produces the sporophytic phase of the plant.

Zygote
Greek
zugon to yoke, pair
A cell formed by the union of two gametes.

Zymurgy
Greek
zym- leaven
-ourgos work
The branch of chemistry that deals with the process of fermentation.

Common Prefixes

a- no, absence of, without, lack of, not

ab- off, away from

acere- to be sour

ad- to, a direction toward, addition to, near

aden- lymph gland(s)

aequi- equal, same, similar, even

aer- air, atmosphere, mist, wind

algeis- pain

alkali- (**Latin**) basic, pH more than 7

allos- other, different

alqili- (**Arabic**) ashes (originally from Arabic word al-qali, which means "ashes," and recalls the elements Na [sodium] and K [potassium] left in the ashes of burning wood or plants)

amnion- embryo, bowl, lamb

amphi- on both or all sides, around

an- no, absence of, without, lack of, not

ana- anew, up

andros- male

anemos- wind

angeion- vessel, usually a blood vessel

ante- before or prior to

anth- flower; that which buds or sprouts

anthropo- man; human being, mankind

anti- opposing, opposite, against

apo- away from, off, separate

aqua- water

archae- original, beginning, origin, ancient

artēriā- windpipe, artery

arthr- joint

astros- star

athera- tumors full of pus, like a gruel

atmos- vapor

atri- open area, central court, hall, entrance, or main room of an ancient Roman house

auto- self, same, spontaneous; directed from within

avis- bird

baktron- a staff; rod

baro- weight, heavy; combining form meaning "pressure"

bathy- deep, depth

bi- two, twice, double, twofold

blastos- germ, bud

brakhīōn- upper arm

bronkhos- windpipe

centi- one hundredth

cephalo- (kephalikos) head

chaeto- spine, bristle; long, flowing hair

cheil- claw, lip, edge, or brim

chemo-, khemeia- chemical/alchemy

chlor- the color green, yellow-green, or light green

circum- in a circle; around, about, surrounding

co- to the same extent, degree; together, jointly

com- (con) together, with, jointly; compress, converge

cyano- (kyanos) blue, dark blue

dactylo- finger, toe

de- do or make the opposite of, reverse the action of, undo; from, apart, away

deinos- terrible, monstrous

dendro- tree, resembling a tree

dermat- skin

di- apart, away, from, two

dia- through, across, apart

diploos- double

dis- apart, away from, utterly, completely, in all directions

dys- painful, difficult, disordered, impaired, defective, ill

e- out

ektos- outer; external, out of, out, outside; away from

ēlektron- charge, electricity, dealing with positive and negative charges

en- in, into, inward; within

endo- inside, within

environ- round about; encircle

epi- above, over, on, upon

eu- good, well; true

ex- outside/outward, out of, out; away from

ferrum- iron; pertaining to or containing iron

fibro-, fibr-, fibra- fiber; an elongated threadlike structure

frangere- to break

gamet- husband or wife; to marry

gastr- stomach, belly

ge- earth, world

gen- origin, birth

germen- a bud, offshoot

gravis- heavy, weighty

haima- blood

hēlio- sun

hemi- half

hepta- liver

herba- grass, green crops

heteros- different

holos- complete, whole, entire, all, full

homeo- same, like, resembling, sharing, similar, equal

hydr- of or having to do with water

hyper- above, high

hypo- under, below, beneath, less than, too little, deficient

infra- inferior to, below, or beneath

inter- among, mutually, together, between, among

intra- within, inside

isos- equal, uniform, same, similar, alike

kard- heart, pertaining to the heart

kary- nut, walnut, kernel, nucleus

kata- down, downward; under, lower; against; entirely, completely

kentron- center, sharp point

khondros- granule, cartilage

khromat- color

kinetikos- to move; set in motion

klinein- to lean, sloping

koilos- hollow cavity

kosmos- universe, order

kustis- (cyst) sac or bladder that contains fluid

kyklos- circle, wheel, cycle, rotate

leukos- white, clear, or colorless

lipos- fat

lithos- stone or rock

ly- (luein) to loosen, dissolve, dissolution, break

lympha- clear water, water nymph

magn- great

makros- long, large, great

mala- bad

medius- middle

megas- large, great, big, powerful

melas- (melas) the color black, dark

mesos- middle

meta- between, after, beyond, later

micro- denotes one-millionth of a part

mono- one, single, alone

morph- shape, form, figure or appearance

myco- fungus

myel- (muelos) bone marrow

myo- muscle

necro- death

nephros- kidney

neur- nerve, cord

nervus- sinew, tendon

nom- (nemein) to dictate the laws of, knowledge, usage, order

non- not, lack of

nucula- kernel, little nut

oikos- home, house

oion- egg

or- mouth

ortho- straight, true, correct, right

ōs- mouth

osteon- bone

ovum- egg

pan- all

para- beside, near, alongside

pathos- suffering, disease

ped- foot

per- through, across

peri- around, about, enclosing

petros- a rock, fossil, or stone

phagos- (phagein) to eat, eating

phainein- to show, appear, display; making evident; literally, "to come"

pherbein- to graze

pherein- to carry, bear, support; go

philos- love, fondness for, loving

photos- light, radiant energy

phukos- rock lichen, seaweed

phullon- leaf

phuton- plant

pinein- to drink

plastos- (plassein) something molded; to mold

platus- flat

pneumon- lung, breath

poly- many or much

pro-, prot- before, forward; for, in favor of; in front of

proteros- earlier

pseudes- false

psych- mind, consciousness, mental process

pteron- feather, wing

quadi- four

radi- radiant or radiation energy, wireless transmitter

re- to do something again or go against
rodere- to gnaw
sapro- rotten, putrid, decay
sed- sit
semi- half
sēpein- to make rotten, putrefactive
sinus- hollow
sklero- hard
soma- (somatiko) body
specere- to look at, appearance
spora- seed
staphylo- cluster
statos- standing, stay, make firm, fixed
stereos- three dimensional, solid, firm, hard
stratum- horizontal layer; stretched, spread out; layer, cloud layer
sub- under or below
super- superior in size, quality, number, or degree; exceeding the norm
sus- (sub) below, under, beneath
sym- with, together

syn- joined together, together with
tele- far off, distant
telos- end
thallos- young green shoot
thermos- combining form of "hot" (heat)
thrombo- clot, blood clot
topos- place
trans- across or through
tri- three
trope- bend, curve, turn, a turning; response to stimulus
trophos- (trophein) to nourish, food, nutrition; development
ultra- beyond, to an extreme degree
un- not
uni- same, one
vas- vessel, duct
vena- vein
viva- life, alive
xenos- stranger, different
zoon- animal, animal-like

Common Roots

abdomen   belly, venter

aberrare   deviation from the proper or expected course

abradere   to scrape off

accipiter   hawk

accuratus   done with care

acere   to be sour

acervāre   to heap

activus   to drive, do

āctus   to set in motion

acus (acuere)   to sharpen; needle, point

aden   lymph gland(s)

adip   of or pertaining to fat

aera   counters

aerobe   organism requiring oxygen to live

aesthe   feeling, sensation, perception

aestus   tide, surge

agogos   a leading, a guide

agon   conflict, contest

agulum   to condense, to drive

aion   indefinitely long period of time

aisthesis   feeling

aither   upper air

aitia   cause

akanthos   thorn plant

aktin   ray (as of light), radiance, radiating

albumo   the color white

albus   the color white

aleiphein   to anoint with oil

alere   to nourish

alescere   to come together or grow

alga   seaweed

algesi   pain, sense of pain; painful, hurting

alimentum   nourishment, supplying food

alkali (**Latin**)   basic, pH more than 7

alkyl   alcohol; a monovalent radical, such as ethyl or propyl

alleion   mutually

alligare   to bind

allium   onion, garlic bulb

alqili (**Arabic**)   ashes (originally from Arabic word *al-qali,* which means "ashes," and recalls the elements Na [sodium] and K [potassium] left in the ashes of burning wood or plants)

alter   other

altus   high, highest, tall, lofty

alveus   hollow, belly

am (ampere)   named for Andre Marie Ampere

amalgama   mixture

ameibein   to change

amino   relating to an amine or other compound containing an NH_2 group

ammonia   a colorless pungent gas, NH_3

amnion   embryo, bowl, lamb

ampho (amphoteros)   both, each of two

amplus   large, full

amygdale   almond

analogos   proportionate

ancon   elbow

ane   organic compound containing no multiple bonds

angeion   vessel, usually a blood vessel

angulus   angle

ankhonē   a strangling

annellus   little ring

antara   interior

anth   flower, that which buds or sprouts

anthrankitis   name of a fiery gem

anthropo   man; human being, mankind

aort   lower extremity of the windpipe; by extension, extremity of the heart, the great artery

apatē   deceit

aponeurousthai   to become tendinous

aptare   fit, fitted, suited

aqua water
arakhn spider
arassein to strike
arbor tree
arc bow, arch, or bend
archae original, beginning, origin, ancient
argillos clay
arithmos number
aroma smell (due to sweet smell of benzene and related organic groups)
arteria windpipe, artery
arthr joint
articulus small joint
artificialis not natural, manmade
askarizein to jump, throb
askos bag
astros star
äther etherlike acid
atri open area, central court, hall, entrance, or main room of an ancient Roman house
audit hearing, listening, perception of sounds
augere to increase
auricula ear
aurora dawn
aurum gold
austr south, south wind
auxein to grow
avis bird
awariyah damaged merchandise
axios worthy
axis central
axōn axis
baktron a staff; rod
bar weight, pressure
basid foundation or base
basis fundamental ingredient, foundation
benthos deep; the fauna and flora of the bottom of the sea
beta second letter of the Greek alphabet
bio life, living organisms, or tissue
bios life, living organisms, or tissue
bitūmen a mineral pitch from the Near East
blaedre bladder
blastos bud, germ cell
blepharon eyelid
blōd to thrive or bloom
bol (ballein) to put or throw
bombos booming sound
boreios coming from the north
botah (body) the material frame of humans and animals
botanē fodder, plants
botulus sausage
bov cow
brakhīon upper arm
bredan to breed

bresta to break asunder
brevis brief
bronkhos windpipe
bruein to be full, bursting
bruon moss
bul place for
bulla bubble
buoy to float
bussos bottom
bustus to burn
cadere to fall, die
caecus blind
caelum sky, heaven
caldaria cooking pot
calor heat
calve calf
cambiare to exchange
camoufler to disguise
canālis conduit
cancer crab
candela candle
cani dog
canthus rim of a wheel or vessel
cap catch, seize, take hold of, contain, take, hold
capacitas spacious
capill hairy
cappa cap or cape
carbo coal, charcoal
carbonate to charge with carbon dioxide gas
carota carrot
carpus wrist; that which turns
cartilago cartilage
caud tail
caudex book
caulis stem
cauter heat
cavare to make hollow
cēdere to go
cella chamber
cellula little cell
centrum center
cephalo (kephalikos) head
cer wax
cerebr of or relating to the brain or cerebrum
cernre to separate
cērussa a white lead pigment, sometimes used in cosmetics
cervic stem of cervix
cetu whale
chemo, khemeia chemical; alchemy
chimaira she-goat
chir hand; pertaining to the hand or hands
chore a central and often foundational part, usually distinct from the enveloping part by a difference in nature

chylos juice
ciere to set in motion
circulus to make circular
circum in a circle; around, about, surrounding
cirro hair or wispy
cist to cut
clāvis key (from its shape)
cleave to split or separate
clitellae packsaddle
cloa'cae drain
clupea herring, small fish
coāgulum coagulator
cod a code of laws, a writing tablet; an account book
coelom, (koilomat) cavity
colere to till
commodus to adjust, suitable
communis commons
compose to form, create
conch shell
copula bond or pair
corneus horny
corniculum horn, hornlike structure
corolla small garland
corona crown
cortic bark, rind, that which is stripped off
costo rib
cracian to break apart
cremo, crem to hang; hung, hung up
creper dark
creta chalk
crevace crevice
cropp craw
crum planted with trees
crusta shell, hard surface of a body
cult to care for, to dwell, to inhabit
cumaru tonka bean tree
cumul pile or heap
cumulāre to pile up
currere to coincide
cuspis sharp point, cusp
cutis skin
cutten to separate into parts with or as if with a sharp-edged instrument
cyano (kyanos) blue, dark blue
cygnus swan
cyte (kutos) sac or bladder that contains fluid
daktulos toe, finger, digit
datum something given
decidu to fall off
degrade to impair physical structure
dei god, deity, divine nature
deletes to erase, destroy
deliquiscere melt by absorption of moisture
delo visible, clear, clearly seen; obvious
demos population, people

dendr tree, resembling a tree
dens to press close together
densi thick, thickly set, crowded, compact
denti teeth or tooth
dentis tooth
derm skin
desiccare make quite dry
deterere to lessen, wear away
deuteros second, two in number
diast dilation, spreading
dicho akin to
didumos twins, testicles
diffundere to spread out
digerere to break down
diploos double
diurnus day
diverse differing from another
dold to dull
dominae to rule
domo house, home
doopen to dip
dormire to sleep
dorsalis back
draga to draw, drag
dramein/dromos to run
drum ridge, back; long, narrow hill
ducere to lead, bring, take; to draw or lead
ductus to be hammered out into a tube or pipe; leading or drawing
dunamikos powerful
duodecum twelve
durare to harden; hard growth
dygre dry
eco environment, habitat
efficere to effect
eghe resembling an eye shape
eicere to throw out
eisodios coming in besides, entering
ekdusis to shed or molt
ekithos yolk
elaunein to beat out
ēlektron charge, electricity, dealing with positive and negative charges
elementum rudiment, first principle
eliminat to banish
elleiptikos of a leaf shape; in the form of an ellipse
elongate to make or grow longer
elutron sheath
ēmittere to send out
empeirikos doctor relying on experience alone
enchyma tissue
enkephalos in the head
enteron intestine
entomos cut from two, segmented
equus horse

erbe   herb

erem   lonely, solitary; hermit; desert

ergon   work

erosio   an eating away

estiv   dormancy in the summer

etymon   true sense; earlier form of a word

eurus   a widening; broad, wide

evolut   unrolling

experiri   to try

externus   outward

facere   to do, carry, bear, bring

fecere   make, do, cause, produce, build

ferre   to carry

fibre   an elongated, threadlike structure

flagrum   whip

flēoge   fly

florere   flower; to blossom

focus (fuel)   hearth, fireplace

folium   leaf

foris   outside

formyl: form(ic)   found in ants or relating to ants +
   -yl suffix for organic acid

frangere   to break

fugere   to flee

fungi   performance, execution

furca   a fork

gaia   earth

gastr   stomach, belly

ge   earth, world

gen   to give birth, kind, produce

genitus   born, to bear

gerere   to bear

glene   eyeball

glotta   tongue

glutinare   to glue

glūtīre   to gulp

gnatha   jaw

gnō   to come to known

gnose   to know or learn

gonos   offspring

gradus   step or degree

gradus   walk, step, take steps, move around; walking or stepping

gramma   letter

graphia (graphein)   to write, record, draw, describe

gynous   in relation to a female organ of a plant

haerere   to stick together, cling to

haima   blood

hal   salt

havour   to have

hedron   face

helios   sun

heteros   different

histanai   to place, to stop

homolus   even

hormo   to rouse or to set in motion

hudor   water

hybrida   mongrel offspring

hydr   water

jugare   to join together

kainos   recent

kairon   nut; cell nucleus

kalendae   account book

kalyx   cup

kapnos   smoke, carbon dioxide (CO_2)

kard   heart, pertaining to the heart

karkinos   crab, cancer

karoun   to put to sleep, plunge into sleep or stupor,
   stupefy

karpos   fruit

kata   down, downward; under, lower; against;
   entirely, completely

kele   hernia, tumor

kentein   to prick, puncture

kentron   center, sharp point

keras   horn

kerkos   tail

khartes   map, chart, paper

kheilos   lip

khole   bile

khorde   gut, string of a musical instrument

khorion   afterbirth

khrōma   color

khronos   time

khrūsallid   gold-colored pupa of a butterfly

khumos   juice

kin' dh   to sting, nettle

kine   movement, motion

kinein   to move

kirkos   circle

kirrhos   tawny yellow

klados   branch or spout

klastos   break, break in pieces

kleitoris   clitoris

kleps   to steal

klime   slope

klinein   to lean, sloping

klinikos   pertaining to a bed or couch

klisis   inclination

klōn   young shoot or twig

knēkos   safflower

koiloma   cavity

kokhlias   snail

kokkos   berry, grain, seed

kolkhikon   meadow saffron

kolla   glue

kolon   large intestine

kometes   long-haired

koneion   poison hemlock

konis   dust

kope   oar

kosmos   universe, order

kotuledon   a kind of plant, a seed leaf, a hollow or cup-shaped object

kranion   skull

krater   bowl for mixing wine and water

kreat   flesh

krinein   to separate

kroke   pebble

krustallos   ice, crystal, freeze, icelike

kuhl   essences obtained by distillation

kustis (cyst)   sac or bladder that contains fluid

kyklos   circle, wheel, cycle, rotate

lapar   the soft part of the body between the ribs, hip, and flank; the loin

lātus   wide

legein   word, speech

leipein   to leave

lekithos   egg yolk

libr   balanced, level; make even; weight

ligāre   to tie, bind

ligo   bind, tie

lipo   abandon, to leave (behind)

lite (lith)   stone or rock

locare   to place

luere   to wash, clean

lunar   moon, light, shine

ly (luein)   to loosen, dissolve, dissolution, break

magnes   figurative sense of something that attracts

malacia   softening of tissue

malgama   soft mass

māter   mother

maza   mass, large, amount

mbolon   wedge, peg

megas   large, big, great

melas   black

mensa   table

meros   part

meta   later in time

metallon   mine, ore, quarry; any of a category of electropositive elements from metallum

meter (metron)   instrument or means of measuring; to measure

(meth)ane   an odorless, colorless gas, CH_4

metiri   to measure out

metra   womb

metria (metron)   the process of measuring

migrare   to move

miktos   mixed or blended

minie   mimic, mime; imitate, act; simulation

mittere   to put

mixis   mingling, intercourse

morph   shape, form, figure, or appearance

morpheus   god of dreams

mukēs   fungus

mulgēre   to milk out

myo   muscle

nasus   nose

nautes   sailor

necro   death

negare   say no, to deny

nekros   death, corpse

nephros   kidneys

neur   nerve

nervus   sinew, tendon

nimbus   cloud

noct   night

nom (nemein)   to dictate the laws of, knowledge, usage, order

nosia   disease

och   fixed

ocul   of or relating to the eye

odontos   tooth

oidēma   a swelling

oikos   home, house

optic   eye, optic

orexis   appetite

otic   state or condition of; condition of being

oxo   oxygen

oxus   sharp

oxy   pungent, sharp

parare   to make ready

particula   a very small piece or part; a tiny portion or speck

pathos   feeling, sensation, perception; suffering, disease

pectin   comb

ped   foot

pendere   to hang

peps   digestion

*pestis (**Latin**)*   plague, pestilance

petere   to strive

phage   to eat

phagei   to eat

phagos (phagein)   to eat, eating

phana   speech

pharynx   throat

phase   a stage

phatos   speech, spoken

phile   one who loves or has a strong affinity or preference for

phonos   voice

phore   bearer, carrier

phoreus   bearer

phoros   being carried, bearing

photos   light, radiant energy

phragma   fence

phren   diaphragm, midriff, heart

phuein   to grow

phullon   leaf

phusan   to blow
phusis   nature
phuton   plant having a (specified) characteristic or habitat
phyein   to grow
phyte   plant
pithecus   ape, apelike creatures
plasm (plassein)   to mold or form cells or tissues
plassein   to form
plastos (plassein)   something molded (to mold)
plexus   an embrace
pnea   breathing or breath
pneumon   wind, breath
pnion   breathing or breath
pod   foot
poiein   production, formation; to make
pole   either of two oppositely charged terminals
pollere   to be powerful
ponere   to put together
potent   power; to be able
praktikos   practical
premere   to press
proktos   anus
pteron   feather, wing
ptilon   plume
pūr   fire
pyge   rump or buttocks
pyle   gate
qalib   shoemaker's last
ramus   branch
reciepere   to receive
ren   the kidneys
rhein   to flow or run
riche   rich
rigare   to wrinkle
rocca   rock, stone
rota   wheel
rube   red
saccharon   sugar
safira   to be empty
sauros   lizard
scire   to know
scoli   curvature, curved, twisted, crooked
sectus   to cut
seminare   to plant or propagate (from *semen, seminis* meaning "seed")
sentīre   to feel
sepein   to decay, cause to rot
sepsis   putrefaction or decay
ser   the watery part of fluid
servare   to preserve
sexus   sex
sicca   drying
simulare   to make similar or alike
skeletos   dried body

sklero (sklēroun)   to harden
skopein   see, view, sight, look at, examine
sociar   to join
solvere   to loosen
soma (somatiko)   body
sorbere   to suck
spargere   to scatter or strew; sprinkle
sperma   seed
sphaira   a globe shape, ball, sphere
sphyzein   to throb; pulse, heartbeat
spir   breath of life, breath, breathing
spora   seed
stare   to stand firm
statos   standing, stay, make firm, fixed, balanced
stele   pillar
stella   star
stereos   solid, being of three dimensions
sthenos   strength
stigma   a point, mark, spot, puncture
stillare   to drip or trickle
stingere   to pull
stinguere   to quench
stipare   to press together
stoma   mouth
sumere   to take
summetros   of like measure
sumptotos   intersecting
sustellein   to contract
sylos   a pillar
systema   the universe
taktos   ordered
taxi   arrangement, order; put in order
teg   touch, reach, handle
tekhne   skill, systematic treatment
temnein   to cut
ten   to move in a certain direction; to stretch, hold out
tenere   to hold together
tenuis   thin
terrēre   to frighten
thalpien   to heat
thele   nipple
therapeuein   heal, cure; treatment
thermos   combining form of "hot" (heat)
thorax   breastplate, chest
tomos (temnein)   to cut, incise, section
tonos   tone, stretching, firm
topos   place, spot
tornāre   to round off
toxikos   poison
trahere   to draw
tribuere   to give
tripsis   a rubbing (so named by its first being obtained by rubbing a pancreas with glycerin)

trope   bend, curve, turn, a turning; response to stimulus

trophos (trophein)   to nourish, food, nutrition; development

trudere   thrust

tundere   to beat

tupos   type, model, stamp

unus   one

vacare   empty

vagina   sheath

valere   to be strong

valve   leaf of a door

vaporatus   steam, vapor

variare   to vary

vascul   small vessel

vehere   to carry

vent   come

ventricul   belly

verge   to tend to move in a particular direction

vertere   to turn, turn around

vextus   to be vaulted

vorare   to devour

vore   eat, consume, ingest, devour

weike   pliant

zein   to boil

zoe   life

zoon   animal, animal-like

zuma   leaven, yeast

Common Suffixes

-*a* (**plural**) structure
-*able* capable, be inclined to, tending to, given to
-*able/-ible* capable of
-*ac* pertaining to
-*ad* member of a botanical group
-*ae* plural
-*age (āticum)* (**Latin**) condition or state
-*al* of the kind of, pertaining to, having the form or character of
-*algia* pain, sense of pain; painful, hurting
-*an* one that is of or relating to or belonging to
-*ance* brilliance, appearance, state, quality
-*ancy* condition of or state of
-*androus* man, men, male, masculine
-*angeion* diminutive of "vessel"
-*ant* having the quality of
-*ar* relating to or resembling
-*ary* of, relating to, or connected with
-*ase* enzyme
-*ate* of or having to do with
-*ate* an organism having these characteristics; characterized by having; a derivative of a specific chemical compound or element
-*baros* weight, heavy, atmospheric pressure
-*benthos* deep; the fauna and flora of the bottom of the sea
-*blastos* bud, germ cell
-*cephaly (kephalikos)* head
-*chrome* pigment
-*cide (caedere)* to cut, kill, hack at, or strike
-*cy* state, condition, quality
-*cyst (kustis)* sac or bladder that contains fluid
-*dactylos* finger, toe
-*derm* skin
-*dynia* pain
-*ectasis* expansion, dilation
-*eilema* veil, sheath

-*ekt* outside, external, beyond
-*ella* little, diminutive
-*emesis* vomit
-*emia* the condition of having (a specific thing) in the blood
-*en* to make or cause
-*ence* the condition of
-*ent* causing an action, being in a specific state, within
-*er* one that performs an action
-*ferre* to carry
-*ferrous* bear, carry; produce
-*forma* having the form of
-*fy (ficare)* cause, to become; make, do, build, produce
-*gen* to give birth, kind, produce
-*genus* offspring, kind
-*geny* birth, descent, origin, creation, inception, beginning, race, sort, kind, class
-*gram* something written or drawn; a record
-*graphia (graphein)* to write, record, draw, describe
-*haima* blood
-*haptien* to fasten, join
-*ia* names of diseases, place names, Latinizing plurals
-*ial* (variation of -*ia*) relating to or characterized by
-*ic (ikos)* relating to or having some characteristic of
-*id* state, condition; having, being, pertaining to, tending to, inclined to
-*ide* binary compound; group of related chemical compounds; nonmetal radical
-*ify (ficus)* make, or cause to become
-*il* substance relating to
-*ile* changing, ability, suitable, tending to
-*in* protein or derived from a protein; neutral chemical
-*ine* of or relating to; a chemical substance
-*inferus* below, low
-*ing* the act of or action
-*ion* state, process, or quality of
-*ion (ienai)* to go, something that goes

-ious full of, having the quality of, relating to

-ism state or condition, quality

-ist one who is engaged in

-ite minerals and fossils; component of a part of a body; a part of or product of

-itis inflammation, burning sensation

-ity state of, quality of

-ium quality or relationship; chemical element

-ive performing or tending toward a specific action

-ixation action, process, or result of doing or making

-ize to make, to treat, to do something with

-klastos break, break in pieces

-klinein to lean, sloping

-lin small or little

-lite combining form used in naming of minerals

-lithos stone or rock

-logic talk, speak; speech, word

-logist one who speaks in a certain manner; one who deals with a certain topic

-logos word, proportion

-logy (logos) used in the names of sciences or bodies of knowledge

-lus thing

-ly like, likeness, resemblance

-lympha clear water, water nymph

-lyte substance capable of undergoing decomposition

-mania obsessive preoccupation with something; madness, frenzy; obsession or abnormal desire for

-megaly large

-ment state or condition resulting from a (specified) action

-meter (metron) instrument or means of measuring; to measure

-metria (metron) the process of measuring

-morph shape, form, figure, or appearance

-nom (nemein) to dictate the laws of, knowledge, usage, order

-nosis disease

-odont having teeth

-oid (oeidēs) resembling; having the appearance of

-ol alcohol, chemical derivative

-ole little one

-ologist one who deals with a specific topic

-oma tumor, neoplasm, community

-on a particle

-opsy examination

-or a condition or property of things or persons, person that does something

-ory tending to, serving for

-osis disease or abnormal condition

-ous full of, having the quality of, relating to

-patheia disease, feeling, sensation, perception

-penia reduction, poverty, lack, deficiency

-phagos (phagein) to eat, eating

-pherein to carry

-phile one who loves or has a strong affinity or preference for

-phobos fear

-phyte plant

-plasia (plassein) something molded (to mold)

-plasm (plassein) to mold or form cells or tissues

-plastos (plassein) something molded (to mold)

-plasy growth or development of

-ploid having a number of chromosomes that has specified relationship to the basic number of chromosomes

-pod, -poda, - podos, - pous foot

-ptera feather, wing

-pterux wing

-sis action, process, state, condition

-skopion for viewing with the eye

-soma (somatiko) body

-sphaira a globe shape, ball, sphere

-spora seed, a sowing

-statos standing, stay, make firm, fixed, balanced

-status to come to a stop, to stand

-stoma mouth, opening

-superus higher, upper

-tomos (temnein) to cut, incise, section

-tonia, -tone tension, pressure

-trope bend, curve, turn, a turning; response to stimulus

-trophos (trophein) to nourish, food, nutrition; development

-ula diminutive, little, small

-um (**singular**) structure

-us singular, thing

-y place for an activity, condition, state

-zoan animal

Resources

The American Heritage Dictionary of the English Language, 4th ed. New York: Houghton Mifflin, 2004.

The American Heritage Stedman's Medical Dictionary. New York: Houghton Mifflin, 2002.

Answers Corporation. *Answers.com.* Online encyclopedia, thesaurus, dictionary definitions, 2006.

Bartlett, John, comp. *Familiar Quotations,* 10th ed., rev. enl. by Nathan Haskell Dole. Boston: Little, Brown, 1919; Bartleby.com, 2000.

Biology-Online.org. *Biology Online—Information in the Life Sciences,* 2006.

Bragg, Melvyn. *On Giants' Shoulders: Great Scientists and Their Discoveries from Archimedes to DNA.* London: Hodder & Stoughton, 1998.

Chabner, Davi-Ellen. *The Language of Medicine,* 3rd ed. Philadelphia: W. B. Saunders, 1985.

Columbia University Press. *The Columbia Electronic Encyclopedia,* 6th ed., 2003.

Dictionary of Latin and Greek Words Used in English Vocabulary. Senior Scribe Publications, 2003 to 2006. http://www.wordinfo.info.

Gribbin, John. *The Scientists.* New York: Random House, 2002.

Herron, W. B., and N. P. Palmer. *Matter, Life, and Energy.* Chicago: Lyons & Carnahan, 1972.

Hurd, Charles, and Eleanor Hurd. *A Treasury of Great American Letters.* New York: Hawthorn Books, 1961.

Knowles, Elizabeth. *The Oxford Dictionary of Quotations,* 6th ed. Oxford: Oxford University Press, 2004.

Little, R. John, and C. Eugene Jones. *A Dictionary of Botany.* New York: Van Nostrand Reinhold, 1980.

Merriam-Webster's Collegiate Dictionary, 11th ed. Springfield, MA: Merriam-Webster, Inc., 2006.

Palmer, E. Laurence, and H. Seymour Fowler. *Fieldbook of Natural History.* New York: McGraw-Hill, 1975.

Peterson Field Guide to Rocks and Minerals. New York: Houghton Mifflin, 1998.

Roget's II: The New Thesaurus, 3rd ed. New York: Houghton Mifflin, 1995.

Sagan, Carl. *Broca's Brain.* New York: Random House, 1979.

Sagan, Carl. *Cosmos.* New York: Random House, 1980.

Schuster, M. Lincoln. *A Treasury of the World's Great Letters.* New York: Simon & Schuster, 1950.

Thompson, Ida. *The Audubon Society Field Guide to North American Fossils.* New York: Alfred A. Knopf, 1982.

Whitaker, John O., Jr. *The Audubon Society Field Guide to North American Mammals.* New York: Alfred A. Knopf, 1980.

Wikimedia Foundation. *Wikipedia, the Free Encyclopedia.* 2006. http://www.wikipedia.com.

WordNet 1.7.1. Princeton University, 2001.

About the Author

JOSEPH S. ELIAS is an Associate Professor of Science Education at the Kutztown University of Pennsylvania. He holds a BS in biology (Kutztown University, 1971), an MS in science education (Temple University, 1976), and a doctorate in science education (Temple University, 1989). He has been teaching pre-service secondary education science majors for over 14 years. He also is a university supervisor of secondary education clinical students and teacher interns. Dr. Elias teaches in the graduate school at Kutztown University as well. His graduate courses include methods of research in biology and methods of teaching science to middle and high school students. Prior to teaching at Kutztown University, Dr. Elias taught as an adjunct faculty member in the biology departments of Cedar Crest College and Lehigh Carbon Community College.